Volume 50 March 1988 Pages 1–117

NESTING WEIGHTS, EINSATZGEWICHTE AND PILES À GODETS:

A CATALOG OF NESTED CUP WEIGHTS IN THE EDWARD CLARK STREETER COLLECTION OF WEIGHTS AND MEASURES

by

Ellen Zak Danforth

Medical Historical Library
Yale University
New Haven

TRANSACTIONS
The Connecticut Academy of Arts and Sciences

Published for the Academy by

ARCHON BOOKS HAMDEN, CONNECTICUT

First published for the Academy by
Archon Books, The Shoe String Press, Inc.
Hamden, Connecticut 06414

The paper used in this publication meets the minimum requirements of American National Standard for Information Sciences—Permanence of Paper for Printed Library Materials, ANSI Z39.48-1984.

∞

Library of Congress Cataloging-in-Publication Data

Danforth, Ellen Zak, 1954-
Nesting weights, einsatzgewichte and piles a godets : a catalog of nested cup weights in the Edward Clark Streeter Collection of weights and measures/ by Ellen Zak Danforth.
p. cm.—(The Connecticut Academy of Arts and Sciences transactions ; v. 50, p. 1-00)
"March 1988."
Bibliography: p.
Includes indexes.
ISBN 0-208-02220-1
1. Streeter, Edward Clark, d. 1947—Scientific apparatus collections—Connecticut—New Haven—Catalogs. 2. Weights and measures—Private collections—Connecticut—New Haven—Catalogs. 3. Yale Medical Library. Historical Library—Scientific apparatus collections—Catalogs. I. Streeter, Edward Clark, d. 1947. II. Title. III. Title: Edward Clark Streeter Collection. IV. Series: Transactions of the Connecticut Academy of Arts and Sciences ; v. 50, p. 1-00.
QC81.5.U62N483 1988
530.8—cd19 87-33492
CIP

Cover: "The Weightmaker", an original engraving (20cm x 15cm), by Christoff Weigel, 1698, in The Edward Clark Streeter Collection of Weights and Measures, Yale Medical Historical Library, New Haven, Connecticut.

TABLE OF CONTENTS

INTRODUCTION

Dr. Edward Clark Streeter was fascinated by the evolution of man's scientific curiosity. He began collecting weights and measures in 1923 and based his collecting on the belief that metrology is the foundation of all science. The Streeter Collection of Weights and Measures at Yale University's Medical Historical Library is the culmination of one man's lifelong quest to document science and history.

Although Streeter knew very little about weights and measures when he started collecting them, he was a zealous collector driven by, as he said, "that inner compulsion to acquire and accumulate."[1] It was this instinct, more than anything else, that enabled Streeter to gather together what is one of the most comprehensive metrological collections in the world.

The several thousand artifacts encompass a broad range of material: photographs, metrological ephemera, a library of literature from the sixteenth century to the present, and of course, the scales, weights, and measures of volume and length.

The weights of the ancient classical world are the most important holdings because nearly every type is represented and they are all in good condition. Separately, the Egyptian, Babylonian, Assyrian, Greek, and Roman weights each form an almost complete series.

Other important holdings include the largest collection of French *poids de ville*, outside France or Belgium, and a collection of nested cup weights, which at eighty pieces, is one of the most important museum-owned collections anywhere.[2] The subject of this catalog is this particular group of nested cup weights in the Edward Clark Streeter Collection of Weights and Measures.

When Streeter presented his collection to the Medical Historical Library in 1941 it was not cataloged. He was made Curator of Museum Collections and continued to collect weights and measures for the library until his death in 1947, but left no formal record of his acquisitions and had made only provisional listings of the main objects in his collection. Streeter left the library the awesome responsibility of caring for a collection that was barely inventoried.

Subsequently, several attempts were made to catalog the collection. Dr. Bruno Kisch worked for ten years with it, jotting down notes on hundreds of 5 x 8 inch file cards. Instead of a catalog, however, he produced *Scales and Weights: An Historical Outline* in 1965. Two years later, still hoping to publish a catalog as a "companion" to *Scales and Weights*, Kisch died.

During the 1970s, Edward Clark Streeter's son, John W. Streeter, as Honorary Curator of the Collection, secured the consultative services of French metrologist François Lavagne who, for a brief period, worked intensively with half of the collection, adding substantially to the notes made by Dr. Kisch.

By 1980, however, the collection had been relocated within the Library several times. It had grown to encompass a broad range of material and the "preliminary catalog" of the collection involved notes in several hands and various languages. At this point I became associated with the collection and began ordering and supplementing these notes. A catalog of the items in the collection is now available in the Medical Historical Library at Yale University.

This monograph on the nested weights group is being produced for several reasons. The nested weights represent a portion of the entire Collection that is particularly strong in quality and quantity yet is not too large from a logistical standpoint. The group can stand on its own aesthetically as well.

The visual appeal of the nested weights group is important to a general audience. But for scholars the need for a systematic presentation of data on a cohesive group of nested weights is also apparent. G. M. M. Houben's *2000 Years of Nested Cup-Weights*, published in 1984, is the latest, most complete reference on the subject. But few, if any, catalogs on nested weights exist, thereby frustrating scholars' attempts to compare and contrast data in nested weights collections all over the world.

This catalog of the nested weights group in the Streeter Collection of Weights and Measures will contribute towards a pool of metrological data and encourage the production of similar publications. Historians, metrologists, and collectors all need more catalogs of weights and measures. Without this kind of primary resource material, the important work of reconstructing metrological history is left to hearsay.

The scope of the nested cup weights group in the Streeter Collection is broad and cohesive. Sets range from 64 marks to 4 ounces. Nested cup weights from Nuremberg (*Einsatzgewichte*), England (*nesting weights*), France (*piles à godets*), Spain, Italy, and

Russia are all represented, as are the binary, decimal, duodecimal, and Système Métrique Usuel numerical systems.

The history of the acquisition of this group is not well documented. It is known that Edward Clark Streeter procured, through purchase, most of the nested weights. He bought them primarily from dealers in New York City and Europe over a period of approximately ten years from 1937–1947.

Bruno Kisch, who traveled extensively in North America, Europe, and the Near East to study other metrological holdings for his book *Scales and Weights*, was eventually smitten by the same collecting bug that had bitten Streeter, and bought many weights and measures himself. Although he kept most of his finds, he donated some things to the Streeter Collection. Two of the entries are documented as Kisch donations (Catalog Nos. 52 and 55) and the several that have no donor documentation at all could very possibly be from Kisch's collection as well.

Two other individuals, Clements C. Fry and Mrs. Rodney Chase also contributed to the collection. Two sets of nested weights, originally in Fry's possession, were accessioned in 1958 (Catalog Nos. 59 and 63), three years after Fry's death. In 1973, Mrs. Chase donated eleven sets to the Historical Library (Catalog Nos. 6, 9, 11, 14, 27, 29, 33, 37, 39, 45, and 50).

Description of Nested Cup Weights

Since nested cup weights have already been described by Houben, Kisch, Lockner, and others,[3] only a summary, culled from their writings, will be given here.

Nested cup weights (the term "nested weights" will be used interchangeably) are sets of weights, of cupshape, that fit into each other. The largest cup weight, called the "house," weighs exactly the sum of all the smaller cups. The second largest cup weight weighs half the house and the sum of all the remaining cups. This ratio continues from cup weight to cup weight down to the smallest weight which is called a "disk" or "endpiece." The endpiece, a solid, flat-topped weight with sloping sides of the same degree as the cups,[4] weighs the same as the smallest cup-weight. This principle remained fairly constant until the metric system was introduced. After that, the sequence varied.[5]

Nested weights should not be confused with stacked disc weights. Crawforth defines "stacking weights"—also referred to as "telescope weights"—"as weights which rest on top of each other in descending order of weight, and located by a shallow rim around the top of each weight."[6] Stacked weights do not have houses. There is, however, one notable exception to this rule: Russian nested weights from the nineteenth century are not nested cups but stacked disks inside a specially constructed domed house (see Catalog Nos. 65–67).

All nested cup weights in a set (excluding Russian sets) fit precisely one into the other. The rims are flush. The sides of the cups have a slight concave taper to ensure both a good fit and easy release.

This precise fit is important in determining the condition of nested weights. A complete set has all its original cups. Even a set which has all its original cups but is missing only the small disk, is considered incomplete. If one cup does not fit precisely into the other it is probably a replacement made by someone who believed he or she was increasing the value of an incomplete set. In fact, the set is more valuable to scholars without the misleading (or conflicting) information that an extrinsic[7] weight places upon the set. The scholarly value of a set is determined more by the amount of conclusive information one can get from it and less by how complete it is.

Nested cup weights come in a variety of sizes. Some are so big that at 64 pounds they were difficult to carry and some are so small that at 4 ounces they must have been easily misplaced.

The decoration of these weights is as diverse, too. On some, hunting scenes were engraved on the outside of the house and the lid was adorned with mermaids, knights, horses, sea horses, and sea dogs. Simpler designs had plain sides banded only by engraved, concentric lines, and unadorned lids. Some, like those in England, were not only lidless, but without houses as well.

The Streeter Collection has examples of both simple and elaborate design. It does not, however, have a set made out of silver, another coveted attribute, for it was very rarely used as a material for the manufacture of nested weights.

Bronze was not commonly used either. Exceptions include the antique Roman nested weights (see Catalog No. 1), those made in the Middle Ages, and some sets made as offical standards. The most commonly used material for nested weights was brass. Except for Catalog No. 1, all nested weights cataloged here are made of brass.

Sometimes it is unclear whether a set is made of brass or bronze. Often the patina is responsible for this confusion since copper is the basic alloy for both metals. Fortunately, however, it is now possible to determine rapidly and with relative ease, the precise composition of metal objects using nondestructive techniques. Modern machine analysis of this kind is done in the larger museum conservation studios or in university chemistry laboratories.[8]

Historical Significance of Nested Weights and the Problems of Interpretation

The most common use for nested cup weights was commercial. Rare or extremely elaborate sets were not usually made for ordinary merchants and consumers but for privileged persons[9] or for the purpose of weighing precious metals or gems.

The usage types for the nested cup weights described in this catalog are "commercial," "bullion," "coin," and "inspector's set." Commercial nested cup weights were used to weigh household commodities bought and sold or exchanged by ordinary citizens. Bullion nested cup weights are weights which were used for weighing such precious metals as gold and silver. Nested cup weights used to weigh coins—coins being made of gold or silver—were also similarly precise.

Terms such as "precision weights," "troy weights," "value weights," and "banker's weights" have been used, often interchangeably, and by various people (author included) to describe both bullion- and coin-type nested weights. The confusion over terms is related to the fact that the weighing of money (coins) has always been related to the weighing of gold and silver.[10]

Nested cup weights that were used as official standards are relatively rare and often uniquely enhanced either by design or in the manner that they have been kept. It was not uncommon to keep the standard set in a custom-made box or case. National standard nested weights—such as the *Pile de Charlemagne* from France—are often renowned objects in and of themselves.[11] Local standard weights, although less well known, fulfilled the same purpose; they were the "yardstick" by which all other weights of the realm were measured.[12]

The Streeter Collection owns no standard sets of nested cup weights. It does have, however, an "inspector's set" (Catalog No. 80), a type which was used by official sealers—also called adjusters—to check commercial and other types of weights for accuracy. Inspector's sets are not the same as standards, but the distinction is subtle and merits some more explanation.[13]

According to Gurley's *Handbook for the Use of Sealers of Weights and Measures,* published in several editions in the early decades of the twentieth century, the accuracy of a sealer's "standards and apparatus" are of great importance:

In order that the sealer may be sure of his tests, he should compare at intervals, possibly monthly, his working standards with his office standards, and periodically the State Sealer should inspect the local sealer's outfit and make the necessary alterations or corrections, or have them sent to a responsible manufacturer of sealer's apparatus for repairs, and to the State Department of Weights and Measures or the National Bureau of Standards for verification.[14]

"Working standards" and "office standards" are of the same type as the inspector's set described in this catalog. The standard weights kept by the State Sealer would be equivalent to the local city standards of Europe. The National Bureau of Standards keeps the national standards for the U. S. They are the modern American equivalent of France's *Pile de Charlemagne.*

Other known types of nested weights—examples not included in this catalog—which reflect their specific uses were apothecary-, carat-, corn-, person-, and tolerance-type nested weights.[15] Tolerance weights were used to determine the tolerance, or variation, that was acceptable when checking the accuracy of either weights or balances.[16]

Regardless of its usage type, whenever a nested cup weight was checked for accuracy by an official, that official left his mark or stamp in the weight to show when and/or by whom it had been approved. These marks are called "verification stamps" and are usually located on the lids of houses, or on the bottom or rims of the inner cups.

Although this was an important activity and one that merits the careful scrutiny of historians, the rules for marking nested weights were always changing. A set covered with many marks is difficult to decipher. (See, for example, Catalog No. 36.) Since nested weights were often made in one country and used in another, as is the case with the Nuremberg nested weights, a "country letter"—the first letter of a country or city—was sometimes punched into the lid of the house of a set bound for export. This was not technically a verification (since the mark was made by the maker and not the official adjuster) but a designation.

Also, the marking at the time of verification was not consistent. Sometimes the year was used. Some weights were stamped with the coat of arms or the name of the city in which it was used, and in other examples, the name or initials of the official adjuster were used. Because of these inconsistencies, and the fact that verification stamps as a body of data remain virtually uncharted,[17] this catalog will not attempt to identify the meaning of each and every verification stamp. However, it will specify which marks are verification stamps, which are country letters, and which are maker's marks. In Catalog No. 29, for example, all three appear on the lid.

Maker's marks have been found on nested cup weights as old as those from ancient Rome.[18] Most maker's marks however, appear on nested weights made in Nuremberg from the fourteenth to the nineteenth centuries. These marks, sometimes called "master signs," were like signatures of the individuals who made the nested weights. Even nested weights made in the twentieth century by companies instead of individuals, such as W. & L. E. Gurley Co. (see Catalog No. 80), bear the name of the individual—or individuals—who started the company.

The production of nested weights was usually a group effort. Whether by a family or a guild, the manufacture of nested weights was a complicated procedure requiring the specialization of several individuals. In Nuremberg, for example, "formers" made the molds, "Drechslers" operated the lathes and someone else, no doubt, was responsible for sales and export. Apprentices, too, were surely employed. More accurately then, the maker's mark refers to the shop where the weights were produced, and not necessarily to the weights' designer or fabricator.

It is for this reason (and for others to follow) that this catalog will make little attempt to identify maker's names from marks. Only a description of the mark will be given. Resources for matching marks to names already exist.[19] Conclusive interpretation of maker's marks can only be realized when enough examples and data describing the context in which they appear are carefully reviewed. Because this may involve examination of nested weights not a part of the collection it is a task outside the scope of this catalog.

Like verification marks, then, the problem of interpreting maker's marks is multifarious. One important aspect of this problem is that of dating.

In the city archives of Nuremberg, the individual names of *Rotschmiedmeisters* (master coppersmiths), their specialties, their marks, and the date they passed their master exam are registered.[20] These records cover a period of 500 years from 1375 to 1875.[21] Most of the makers and their marks, however, were recorded in the sixteenth to eighteenth centuries, during the zenith of nested weight production, when Nuremberg had a monopoly on the industry.[22] Virtually all of the nested cup weights used in Europe at that time were made in Nuremberg by coppersmiths who became "certified" masters under a system that required the passing of a special exam (the creation of one or more "masterpieces") as a prerequisite for Rotschmied guild membership.

The standards of the guilds were so high that the mastersigns or maker's marks became more than just signatures: they became symbols of quality. Like the status symbols of designer's logos today, certain maker's marks were more prestigious than others. Once the business was established, the master inevitably had less and less to do with the production of each individual item that came out of his shop. His name became a "rubber stamp" of approval for what was essentially a group effort. This led to such practices as the purchasing of another's name or the use of more than one mark per maker.[23] Also, the similarity of many marks suggests that some were copied.

By the eighteenth century the guilds were starting to lose their influence and for a time many makers did not use maker's marks at all.[24] This gradual decline and the fact that, whether intentional or not, variations in marks occurred over time—maker's names and marks were handed down consanguineously or through marriage by generations of weightmakers—is responsible for some of the difficulties in dating nested weights.

Other factors must also be considered when dating nested weights. Reference has already been made to material, design, and decoration. All of these are important in determining when and where a set was made. Unfortunately, the Nuremberg archives do not record the styles of the various makers.[25]. In general, though, the simplest styles from Nuremberg and elsewhere were made during the early (through the fifteenth) and later (nineteenth and twentieth) centuries, and the more ornate styles were made during the sixteenth, seventeenth, and eighteenth centuries.

City records, maker's marks, and dates of master exams not withstanding, it is still difficult to pinpoint the date of a nested weight's "birth." Although specific dates may appear on nested weights often they are associated with verification and not manufacture. In this catalog, the general period of use is indicated by century.

Mathematics of Nested Weights

Material, design, decoration, verification marks, and maker's marks are not the only attributes of a nested cup weight that can be described objectively. Probably the most vital of a nested weight's statistics is its mass quantity and unit of measurement. This, more than anything else, helps identify the set's usage type, place of use, and period of use.

Coins, for example, were measured in Hungarian ducats, Rheinish guldens, French crowns, Spanish pistols and reals, and English shillings. Bullion was measured in marcs, ounces, gros, or deniers in France (in Germany, the mark), and by the troy pound in England. Commercial goods were measured by the avoirdupois pound in England, the livre in France, and the pfund in Germany. Other examples are pharmaceuticals, which were measured by the troy ounce and fractions thereof (drachmas), or diamonds and pearls, measured in carats. In England, people were weighed in "stones."

Every country, and often different cities within countries, had their own mass equivalent of, for example, the "pound." There are dictionaries that present these variations of theoretical—or ideal—mass in grams.[26] The English pound of 16 ounces (avoirdupois), for instance, was 453.6 grams, the Paris 16-ounce pound (livre) was 489.5 grams; the English pound of 12 ounces (troy) was 373.2 grams, the Milanese 12-ounce pound (libbra) was 326.8 grams. Zurich had an 18 ounce pound of 528.6 grams.[27]

Variations in mass could be so slight that some nested weights are virtually impossible to place precisely. Other factors, such as numerical systems and the arithmetic relationship between the cups in a set, have to be considered.

Regardless of mass, nearly all Nuremberg nested cup weights were made according to the binary system[28]—a system based on an arithmetic breakdown by twos: the cups in a typical 32-pound binary set might be 16-, 8-, 4-, 2-, 1-pound and 8-, 4-, 2-, 1-, 1/2-, 1/4-, 1/8-ounce cups, and 1/8-ounce disk, respectively.

The 12-ounce pound, with its Roman origins,[29] is characteristic of the nested weights from Italy. They were sometimes constructed according to the duodecimal system; the cups in a typical 24-ounce set might be 12, 6, 3, 2, 1, 1/2, 1/4, and 1/4 ounces, respectively. However, since several regions in Northern Italy were at different times under the political influence of other European countries, nested weights from Venice or Milan, for example, were often made according to the binary system.[30]

Sometimes the mass equivalents in any one locale changed over time. Most dramatic was the introduction of the metric-decimal system in the nineteenth century which today provides metrologists with a continuum marker that is especially clear. In France the system was made compulsory in 1840, and all but the Anglo-Saxon world followed soon afterwards.[31] A typical 500-

gram nested weight—with an arithmetic breakdown of 500, 200, 100, 100, 50, 20, 10, 10, 5, 2, 1, and 1 gram, respectively, per cup—is known to be post-1840.

Variations on the metric system usually refer to the period of transition that preceded its legal adoption. These nested weights are called "pre-decimal" or "transitional pieces." A particular pre-decimal system, called Système Métrique Usuel, was officially adopted in France from 1812 to 1840. It was a binary system based on a pound (livre) of 500 grams. Both livre and gram equivalents appeared on the cups in an arithmetic breakdown of 1 livre/500 grammes, 4 onces/250 grammes, 2 onces/125 grammes, 1 once/62.5 grammes, 4 gros/31.2 grammes, 2 gros/15.6 grammes, 1 gros/7.8 grammes, 1/2 gros/3.9 grammes, 1/4 gros/1.9 grammes, and 1/4 gros/1.9 grammes, respectively per cup.[32]

The last factor that complicates the accurate identification of nested cup weights is the problem of adjustment. Since many of the Nuremberg nested weights were made for export, each buyer's mass-units and "size-wishes"[33] had to be figured into their construction. Once the nested weights reached their final destination, local authorities could then make accurate adjustments. Also, the life span of many nested weights surely included useful residence in more than one place; if the mass equivalents varied in these different areas, accommodating adjustments would have had to have been made.

Adjustments made at the time of construction were done by changing the form of the lid and/or house. Unadorned or only slightly adorned lids and low, squatty houses are good examples of this type of adjustment.

Adjustments done *after* construction are more varied. Both minor changes—like filing—and major changes—like routing—were common post-construction adjustment procedures. Other examples whereby excess weight was removed for adjustment purposes include the breaking off of pieces on the lid of the house. These, of course, should not be confused with damaged weights.[34]

The addition of weight was another kind of post-construction adjustment. If the mass was too low, the weights could be adjusted by putting molten lead on the bottom of the cups.[35] Another method was to add a piece of brass to the lid in an inconspicuous spot. Sometimes, as is the case with Catalog No. 15, an actual coin weight was soldered onto it.

All of these adjustment techniques were designed to render the sets of nested weights accurate in relation to the theoretical mass of any city's weight standard. However, despite adjustment (and even verification of mass accuracy by official adjusters), discrepancies between actual mass and theoretical mass commonly occur. Both erosion of metal—either from constant use or abuse—and especially errors in measurement account for most of the discrepancies.

Conclusion and Acknowledgments

It must be emphasized that this catalog highlights only the objective data that can be gleaned from looking at nested cup weights. Specifically, eighty nested weights in the Streeter Collection of Weights and Measures are described. Much of the historical context in which they were made and used will be reconstructed by the reader. Thus, I hope that this book will generate more questions than answers. If, within a short time of its publication, it becomes filled with addenda and question marks, I will consider it a successful effort.

It will be twice as effective if it inspires another catalog of another collection of nested weights, for these collections are meant to be studied, and compared, and should be made as accessible as possible.

Accessibility has been a recurring theme in the history of the collection. The Historical Library was founded by, and designed to accommodate the collections of three men, Drs. Harvey Cushing, John Fulton, and Arnold Klebs. Before the library was built, however, Dr. Streeter, a good friend and colleague of these men, was convinced by them that his collection of weights and measures would be an important addition to the library. Fulton wrote to Klebs on August 16, 1937: "Harvey Cushing, without much warning, organized a trip to . . . see Streeter in the hope of getting his books and his collection of weights and measures for our trinitarian scheme. I think he has succumbed."[36]

After his visit to see Dr. Streeter, Cushing wrote of his enthusiasm for the collection and alluded to why he felt it belonged in a library devoted to the history of science and medicine:

> [Streeter] showed [me] his collection of weights and measures and books pertaining thereto, at all of which . . . I simply gasped. So far as I know, no one else has taken an interest in collecting these things, and he has been picking them up here and there for the past twenty years. . . . No doubt therein lay the beginnings of exact measurement which not only was important for commerce but must also hold in some way the kernel of early science.[37]

More than thirty years later, John W. Streeter, son of Edward Clark Streeter, had this to say about accessibility:

> Crowded almost to the point of jumbling in [the] Sterling Hall of Medicine, much of the Collection is displayed in one small room . . . near the rotunda honoring Dr. Harvey Cushing, whose friendship with Dr. Streeter was a factor in the location of the Collection in the Medical School. Recent checks in London, Paris, New York and Washington prove that, modest as the premises are, the number of items actually on display is greater than in any other museum.[38]

It is the institution's responsibility to both care for and share its cultural resources. From a public vantage point, this responsibility is accepted the moment it accepts the goods. Unfortunately, this is easier said than done. Collections are not inherently income-producing properties, but the opposite. Nevertheless, the responsibility should be met with the best of all possible intentions, and expectations should be high.

Periods of high expectation have generated much that is good for the collection: improved storage and care, better organization, greater visibility, and quality research. Today the Streeter Collection is more safely accessible than ever before.[39]

This catalog itself is a result of high expectations and greater accessibility. High expectations on the part of Dr. Edward Clark Streeter, Dr. Bruno Kisch, and John W. Streeter are to be acknowledged. It is to their efforts that this monograph is dedicated.

The high expectations of the library staff, present and past, specifically those of Ferenc Gyorgyey, Bella Berson, Betty Feeney, and Madeline Stanton are also to be acknowledged.

Michael Coleman and David F. Musto, M.D., both of the Medical School, have each had high expectations for the Streeter Collection. Their encouragement has been invaluable.

I would also like to acknowledge my husband, David L. Danforth, for his support and for taking the pictures in this catalog. David suggested that I include a photograph for each entry, not just a selection. His extra effort on my behalf means that the catalog is far more complete.

Finally, because of the improved accessibility to the collection, many scholars—men and women who have visited the collection and/or corresponded with me—have left their mark, and enriched the collection with their knowledge. Most of them have, in one way or another, contributed to the preparation of this book. I alone, however, bear responsibility for any errors it may contain.

I would like to acknowledge, by name, Gerhard Houben, Michael Crawforth, and Gary Batz, for they have not only contributed so much to current metrological theory, they have also taught me a great deal. Their genuine, enthusiastic interest in the subject was infectious and their knowledge was imparted with such a degree of respect and integrity that it was a pleasure to be their student. I sincerely hope that their association with the Streeter Collection continues indefinitely.

Notes

1. Streeter, "Comment on Collecting," 1.
2. Houben, *2000 Years of Nested Cup-Weights*, 3; Houben to the Historical Library, 24 June 1983. Although Houben mentions seventy-five nested cup weights in the Streeter Collection, there are eighty cataloged here. This is because I added five more sets that are actually part of several coin scale and weight boxes in the collection. Normally the separation of integral parts of a single artifact is frowned upon. However, I do it here only on paper and with the understanding that some nested cup weights being studied today have already been separated—either willfully or inadvertently, but nevertheless permanently—from their "parent" boxes. Although "orphan" sets have value in and of themselves it is important to know that they once had parents. The documentation under "Usage Type" for weights in the Streeter Collection will specify, for example, if the set is, or once was, from a box of coin scales and weights. I hope this will help others identify "orphans" more accurately, too.
3. Houben, *2000 Years of Nested Weights Cup-Weights*; Kisch, *Scales and Weights: An Historical Outline*; Kisch, *Gewichte und Waagemacher im alten Köln*; Lockner, *Die Merkzeichen der Nürnberg Rotschmiede*.
4. I am grateful to Gary Batz (personal communication, October 1986) for suggesting this.
5. *Equilibrium*, 1978, 17.
6. Crawforth, *Handbook of Old Weighing Instruments*, 86.
7. Crawforth to author, 29 August 1986. I am grateful to Michael Crawforth for suggesting this term to identify cups in a set that are not original to it.

8. I am grateful to Jon Eklund, Curator in the Division of Physical Sciences at the Smithsonian Institution, for mentioning this to me and for reiterating the importance of using nondestructive methods. He correctly states that if collectors and curators cannot avail themselves of these services, no tests should be made at all.
9. Houben, *The Weighing of Money*, 46.
10. Crawforth to author, 21 April 1985. I am grateful to Michael Crawforth for pointing out the difficulty in using the term "precision" for describing bullion weights. In addition to the confusion generated by the relationship of coins to gold and silver is the use of the term "precision" for weights used in the scientific laboratory. As far as I know, however, no nested weights were ever intended for laboratory use.
11. Houben, *2000 Years*, 4.
12. As an aside, an interesting article on the subject of standards is "The Greek Metrological Relief in Oxford" by Eric Fernie, in *The Antiquities Journal* Vol. LXI (Part II), 1981: pp. 255–63. The function of this relief, now in the Ashmolean Museum, is thought not to have been for the practical purpose of comparing or checking measures but rather as a symbol for the standard, used, perhaps, at the entrance to a room devoted to the control of weights and measures.
13. I am grateful to Gary Batz (personal communication October 1986) for bringing this problem to my attention.
14. Gurley, W. L. & E., *A Handbook for the Use of Sealers of Weights and Measures*, fourth edition, 9–10.
15. Houben, *2000 Years*, 4.
16. Ibid., 15.
17. Ibid., 67–68.
18. Ibid., 47.
19. Lockner, *Die Merkzeichen*; Houben, *2000 Years*; Stengel, "Die Merkzeichen der Nürnberger Rotschmiede."
20. Houben, *2000 Years*, 47–49; Lockner, *Die Merkzeichen*, 10–17; Stengel, "Die Merkzeichen," 107–14.
21. Houben, *2000 Years*, 47.
22. Kisch, *Scales and Weights*, 126.
23. Houben, *2000 Years*, 49.
24. Lockner, *Die Merkzeichen*, 11.
25. Lockner, however, suggests that several makers are known because the style of their candlesticks or gun barrels is very distinctive. Ibid., 10.

26. See, for example, Doursther, Forien de Rochesnard and Zupko.
27. See Houben, *2000 Years*, 82, for a table summarizing the mass units of the most important countries and cities in Europe.
28. Ibid., 17.
29. Skinner, *Weights and Measures*, 66.
30. Houben, *2000 Years*, 17. (See, for example, Catalog No. 35.)
31. See Kisch, *Scales and Weights*, 21, for a list of countries and the respective dates the metric-decimal system was adopted in each.
32. Houben, *2000 Years*, 22. (See, for example, Catalog No. 51.)
33. Ibid., p. 44.
34. Ibid., 45. (See, for example, Catalog No. 34.)
35. Ibid. (See, for example, Catalog No. 76.)
36. Yale University, *The Making of a Library*, 43.
37. Ibid., 42.
38. Streeter, "The Edward Clark Streeter Collection of Weights and Measures at the Yale Medical Library—One of Yale's Least Known Bests," 5.
39. See Danforth, "The Streeter Collection of Weights and Measures: A New Resource for Metrologists" for a more complete account of the history of the collection.

LEGEND

(key to individual entries)

Note:
1. / means "and/or" when referring to two possible interpretations.
2. Missing cups are in [square brackets].
3. Extrinsic cups are in {braces}; verification stamps on extrinsic cups and any other information about extrinsic cups are also in {braces}.
4. Country and city letters and mass quantity and unit marks are in (parentheses).
5. The catalog entries are grouped primarily by place of manufacture in descending order by size. With exceptions, the entries are listed in chronological order.

Catalog No.: A consecutive number used only in this catalog. All references (e.g., in the indexes) to nested cup weights cataloged here will use this number, not page numbers.

Collection No.: The Streeter Collection number devised by Bruno Z. Kisch. (This is the object's primary number used by the Historical Library for identification purposes.)

Usage Type: "Commercial," "Bullion," "Coin," "Inspector's Set," or "Unknown."

Century: Indicates general period of use.

Place of Manufacture: Country where made (city, if conclusive); country or city letter. Country or city letters are usually located to the left of the locker.

Place of Use: Country where used (city, if conclusive); Country or city letter. Country or city letters are usually located to the left of the locker.

Mass Quantity: In the language of its provenance, the name of the mass quantity and unit. (The sum of the mass of all cups plus the house.) Mass quantity and unit marks are located on the lid, usually to the left of the locker, or sometimes on the "arm."

Theoretical Mass: Ideal mass of a single unit from the particular country or city, in grams.

Numerical System: "Binary," "Duodecimal," "Decimal," "Pre-decimal," or "Système Métrique Usuel."

Numerical Breakdown: Mass quantity and unit of the entire set, the house, and all cups in descending order. Mass quantity and unit marks (in parentheses) inside the house and cups are usually located on the rim or on the bottom.

Actual Mass: The current mass of the house (if present) and all cups in descending order, in grams. Missing cups are not included. For complete sets of a single unit (e.g., 1 pound or 1 mark) or less, the total actual mass is also given.

Maker's Mark: Description of the mark. Maker's marks are located to the right of the locker except where noted.

Verification Stamps: Drawings of the stamps.

Adjustments: Post-construction adjustments only. "Additions" or "subtractions" of metal. Type of material, location, and method of adjustment, if unusual.

Form, Lid: Descriptive terms used are: "w/handle" (with handle); "w/o handle" (without handle); "split V 3-arm"; "single-arm"; "domed"; "flat"; "screw-on type"; "lidless"; "unknown."

Form, House: Descriptive terms used are: "waisted" (concave sides); "barrel-shaped" (convex sides); "straight" (true truncated cone-shaped); "low type" (short truncated cone); "high type" (tall truncated cone); "stitched leather" (a case, actually); "open set" (no house); "unknown."

Decoration, Lid: Descriptive terms used are: "handle w/ horses' heads"; "handle w/ crown"; "plain handle"; "knobs w/ mermaids" ("knobs" refer to the piece holding the handle); "knobs w/ bearded men"; "hinge & lock w/ seadogs"; "hinge & lock w/ seadog-seahorse combination"; "simple hinge & lock"; "split V w/seadogs"; "split V w/seadog-seahorse combination"; "simple split V, rounded"; "simple split V, squared"; "plain" (no embellishments at all, except perhaps, incised concentric lines).

Decoration, House: Descriptive terms used are: "roped" (raised, rope-like bands that go around the outside of the house forming parallel segments for other design elements); "plain" (banded by simple incised lines forming parallel segments without any other design element); "geometric configurations" or "pictorial configurations" (design elements inside the parallel segments). The geometric and pictorial configurations are incised, not raised elements. Prominent pictorial elements are itemized.

Dimensions: Exclusive of lid embellishments, handle, and locker, diameter of top x diameter of bottom x height, in centimeters.

Condition: "Complete" or "Incomplete"; Name and number of missing parts. (Terms such as "good," "excellent," "fair," and "poor" and other qualifiers are too subjective and are not used in this catalog.) If there is obvious wear, and it is believed to affect the theoretical mass, such wear will be mentioned. Also, replaced elements and extrinsic cups are itemized.

History: Donor's initials/Date accessioned (ECS = Edward Clark Streeter; BZK = Bruno Z. Kisch; CCF = Clements C. Fry; MRC = Mrs. Rodney Chase).

Comments: If applicable, additional comments will appear here: cf. = compare (compare this to illustrated example of similar description); ref. = reference (published illustration or description of exactly the same object).

THE CATALOG

Catalog No.: 1

Collection No.: CIRo1–8

Usage Type: Commercial

Century: 5th

Place of Manufacture: Roman Empire

Place of Use: ancient Rome (?)

Mass Quantity: unknown

Theoretical Mass: 324.0g per 12 ounce ancient Roman pound

Numerical System: duodecimal /binary

Numerical Breakdown: 2, [1] libra; 6, [3], [2], [1], 1/2, {1/2}, 1/3, 1/4, {1/4}, 1/8 uncia

Actual Mass: 610.4g, 147.8g, 11.9g, {11.8g}, 5.5g, 5.1g, {5.1g}, 2.6g

Maker's Marks: none

Verification Stamps: none

Adjustments: subtractions: routing on the bottom of both 1/2 uncia cups and one 1/4 uncia cup

Form, Lid: lidless

Form, House: open set

Decoration, Lid: none

Decoration, House: none

Dimensions: of largest cup 7.3 x 6.1 x 4.3

Condition: incomplete; 4 missing cups; at least 2 extrinsic cups (see comments); significant corrosion and encrustation

History: ECS/1938

Comments: This group of nested weights is made of bronze. It is unlikely that this is an original set; it is more likely an "artificial set." Ref. Kisch, 1965, p. 126.

Catalog No.: 2

Collection No.: CIN5

Usage Type: Commercial

Century: 16th

Place of Manufacture: Nuremberg

Place of Use: Spain (S)

Mass Quantity: 4 libra (4)/8 marco

Theoretical Mass: 460.5g per Castile pound / 230.05g per Castile marc

Numerical System: binary

Numerical Breakdown: 4, (2), (1) libra; (8), (4), (2), [1], [1/2], [1/4], [1/4] onza or 8, 4 (4M), 2 (2M), 1 (1M) marc; (4), (2), [1], [1/2], [1/4], [1/4] onza

Actual Mass: 923.1g, 460.2g, 230.2g, 115.3g, 57.5g,

Maker's Marks: cloverleaf w/o initials

Verification Stamps:

Adjustments: none

Form, Lid: split V-3 arm w/ handle

Form, House: slightly waisted; high type

Decoration, Lid: simple split V, squared; handle w/horses' heads

Decoration, House: roped

Dimensions: 8.3 x 5.6 x 6.2

Condition: incomplete; 4 missing cups

History: ECS/1947

Comments: Ref. Kisch, 1965, p. 177; cf. Houben, 1984, p. 53, fig. 97; cf. Lockner, no. 71.

Catalog No.: 3

Collection No.: CIN1

Usage Type: Commercial

Century: 16th

Place of Manufacture: Nuremberg

Place of Use: Spain

Mass Quantity: 4 libra (4)

Theoretical Mass: 460.5g per Castile pound

Numerical System: binary

Numerical Breakdown: 4, (2), (1) libra; 8, 4, 2, 1 onza; 4, [2], [1], [1] ochave

Actual Mass: 923.3g, 458.2g, 114.6g, 57.2g, 28.4g, 14.3g,

Maker's Marks: Moore's head w/o initials

Verification Stamps:

Adjustments: none

Form, Lid: split V-3 arm w/ handle

Form, House: straight; high type

Decoration, Lid: plain

Decoration, House: geometric configurations

Dimensions: 8.1 x 5.4 x 6.7

Condition: incomplete; 3 missing cups; handle, locker, and arms replaced

History: ECS/1944–5

Comments: Ref. Lavagne, 1981, p. 150–51; ref. Kisch, 1965, p. 180; cf. Houben, 1984, p. 51, fig. 90; cf. Lockner, no. 119.

Catalog No.: 4

Collection No.: CIN24

Usage Type: Commercial

Century: 18th

Place of Manufacture: Nuremberg

Place of Use: Spain (S)

Mass Quantity: 4 libra (4)

Theoretical Mass: 460.5 g per Castile pound

Numerical System: binary

Numerical Breakdown: 4, 2, 1 libra; 8, {4}, {2}, {1} onza; {4}, {2}, [1], [1] ochava

Actual Mass: 920.5g, 458.0g, 229.6g, {114.7g}, {56.8g}, {28.1g}, {14.1g}, {7.5g}

Maker's Marks: stork w/ initials L. A.

Verification Stamps: none

Adjustments: addition: soldered lead into and onto both house and cups

Form, Lid: split V 3-arm w/ handle

Form, House: slightly waisted; low type

Decoration, Lid: simple split V, squared; handle w/horses' heads; plain knobs

Decoration, House: roped; geometric configurations

Dimensions: 8.2 x 5.8 x 5.9

Condition: incomplete; 2 missing cups; 5 extrinsic cups; damaged hinge

History: ECS/1941–2

Comments: Ref. Kisch, 1965, p. 177, 182; cf. Lockner, no. 1027.

Catalog No.: 5

Collection No.: CIN19

Usage Type: Commercial

Century: 18th

Place of Manufacture: Nuremberg

Place of Use: Switzerland (S)

Mass Quantity: 2 pfund (2)

Theoretical Mass: 476.6g per Arau pound

Numerical System: binary

Numerical Breakdown: 2, (1) pfund; (16), (8), (4), (2), (1) loth; 2, 1, 1/2, 1/2 quent

Actual Mass: 476.1g, 234.6g, 118.8g, 61.1g, 30.4g, 16.0g, 8.2g, 4.5g, 2.5g, 2.5g

Maker's Marks: half moon w/ initials S.I.

Verification Stamps: none

Adjustments: none

Form, Lid: split V 3-arm w/o handle

Form, House: straight; low type

Decoration, Lid: simple split V, squared

Decoration, House: geometric configurations on a non-segmented, non-banded surface all the way around the house

Dimensions: 6.7 x 4.3 x 4.7

Condition: complete

History: ECS/1943, White Plains, New York antique show

Comments: Ref. Kisch, 1965, p. 182; cf. Houben, 1984, p. 56, fig. 116; cf. Schenk-Behrens, 1987, no. 579; cf. Lockner, no. 1524.

Catalog No.: 6

Collection No.: CIN58

Usage Type: Commercial

Century: 18th

Place of Manufacture: Nuremberg

Place of Use: Spain (S)

Mass Quantity: 16 half-ounces (16) or 1 marco

Theoretical Mass: 230.05g per Castile marc

Numerical System: binary

Numerical Breakdown: 16, (8), (4), (2), (1) half-ounces; 2, 1, 1/2, 1/2 ochava

Actual Mass: 230.4g, 116.0g, 58.0g, 29.2g, 14.6g, 7.5g, 3.75g, 1.8g, 1.8g

Maker's Marks: chalice w/o initials

Verification Stamps:

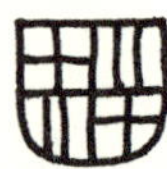

Adjustments: addition; lead soldered into 2 oz. cup

Form, Lid: split V-3 arm w/o handle

Form, House: straight; low type

Decoration, Lid: simple split V, squared

Decoration, House: plain

Dimensions: 4.1 x 2.6 x 2.6

Condition: complete

History: MRC/1973

Comments: Cf. Schenk-Behrens, 1981, no. 399; cf. Lockner, no. 70.

Catalog No.: 7

Collection No.: CIN4

Usage Type: Commercial

Century: 16th

Place of Manufacture: Nuremberg

Place of Use: Spain (S)

Mass Quantity: 1/4 libra or 4 onza (4)

Theoretical Mass: 460.5g per Castile pound

Numerical System: binary

Numerical Breakdown: 4, (2), (1), (1/2), 1/4, [1/8], [1/8] onza

Actual Mass: 115.5g, 57.7g, 28.8g, 14.4g, 7.3g,

Maker's Marks: cloverleaf w/o initials

Verification Stamps: none

Adjustments: none

Form, Lid: split V-3 arm w/o handle

Form, House: straight; low type

Decoration, Lid: simple split V, squared

Decoration, House: plain

Dimensions: 3.4 x 2.1 x 2.1

Condition: incomplete; 2 missing cups

History: ECS/before 1940

Comments: Cf. Houben, 1984, p. 53, fig. 97; cf. Lockner, no. 71.

Catalog No.: 8

Collection No.: CIN21

Usage Type: Commercial

Century: 18th

Place of Manufacture: Nuremberg

Place of Use: Spain

Mass Quantity: 8 libra (8)

Theoretical Mass: 460.5g per Castile pound

Numerical System: binary

Numerical Breakdown: 8, (4), (2), (1) libra; (16), (8), (4), (2), (1) half ounce; {2}, [1], [1/2], [1/2] ochava

Actual Mass: 1836.0g, 919.8g, 459.6g, 229.2g, 114.2g, 57.4g, 28.7g, 14.4g, {7.1g}

Maker's Marks: split-tail mermaid w/initials P.R.

Verification Stamps: 2 illegible marks and

Adjustments: subtractions: routing inside 2 and 1 pound, and 16 and 8 half ounce cups; additions: on 8 half ounce cup it is partially refilled w/lead; lead addition on lid between the split V

Form, Lid: split V-3 arm w/ handle

Form, House: straight; high type

Decoration, Lid: handle w/ horses heads; simple split V, rounded

Decoration, House: geometric configurations

Dimensions: 10.3 x 6.8 x 7.8

Condition: incomplete; 3 missing cups; 1 extrinsic cup; broken hinge

History: ECS/1943–4

Comments: According to his Collection Notes, Lavagne said this was Portuguese (459.0g per Portuguese pound). Ref. Kisch, 1965, p. 183; cf. Lockner, no. 1528.

Catalog No.: 9

Collection No.: CIN53

Usage Type: Commercial

Century: 18th

Place of Manufacture: Nuremberg

Place of Use: Spain

Mass Quantity: 4 libra (4)

Theoretical Mass: 460.5g per Castile pound

Numerical System: binary

Numerical Breakdown: 4, (2), (1) libra; (16), {8}, {4}, (2), (1) half ounce; 2, 1, [1/2], [1/2] ochava

Actual Mass: 917.1g, 459.7g, 229.7g, {115.1g}, {50.0g}, 28.3g, 14.2g, 7.2g, 3.6g

Maker's Marks: mermaid w/ split tail w/initials P.R.

Verification Stamps:

79

Adjustments: none

Form, Lid: split V-3 arm w/ handle

Form, House: slightly waisted; high type

Decoration, Lid: handle w/ horses' heads; simple split V, squared

Decoration, House: roped; geometric configurations

Dimensions: 8.3 x 5.5 x 6.1

Condition: incomplete; 2 missing cups; 2 extrinsic cups

History: MRC/1973

Comments: The '79' is stamped on the lid to the right of the locker. Cf. Schenk-Behrens, 1984, no. 548; cf. Lockner, no. 1528.

Catalog No.: 10

Collection No.: CIN18

Usage Type: Commercial/ Bullion

Century: 17–18th

Place of Manufacture: Nuremberg

Place of Use: France

Mass Quantity: 32 livre (32)

Theoretical Mass: 489.5g per French pound

Numerical System: binary

Numerical Breakdown: 32, (16), 8 (VIII), 4 (IIII), 2 (II), 1 (I) livre; {8}, {4}, [2], [1] once; {4} (oooo), {2} (oo), [1], [1/2], [1/2] gros

Actual Mass: 7818.0g, 3921.5g, 1961.2g, 980.2g, 488.9g, {245.0g}, {121.5g}, {15.0g}, {7.5g}

Maker's Marks: three chalices w/o initials

Verification Stamps: (see comments)

Adjustments: additions: lead soldered into the livre cups and on the lid

Form, Lid: split V-3 arm w/ handle

Form, House: slightly waisted; low type

Decoration, Lid: handle w/ horses' heads; knobs w/ mermaids; hinge & lock w/ seadogs; split V w/seadogs

Decoration, House: roped; pictorial configurations: hunters, deer, dogs, boar, split-tail mermaids, fleur de lis, trees; geometric configurations

Dimensions: 15.8 x 11.9 x 12.1

Condition: incomplete; 5 missing cups; 4 extrinsic cups

History: ECS/1944–5

Comments: This set was hardly used. There are surprisingly few stamps. However, at the base under the hinge, the square stamp is repeated 4 times. One stamp is illegible. Ref. Houben, 1984, p. 28, fig. 43; ref. Kisch, 1985, p. 178; cf. Lockner, no. 70.

Catalog No.: 11

Collection No.: CIN54

Usage Type: Commercial/ Bullion (?)

Century: 17th

Place of Manufacture: Nuremberg

Place of Use: France/Belgium (?)

Mass Quantity: 2 livre (2)/4 marc (?)

Theoretical Mass: 489.5g per French pound/245.9g per Belgian marc (?)

Numerical System: binary

Numerical Breakdown: 2, (1) livre; (16), (8), (4), (2), 1 demi-once; 2, 1, [1/2], 1/2 gros

Actual Mass: 490.0g, 245.2g, 123.0g, 60.8g, 30.7g, 15.2g, 7.8g, 3.8g, 1.9g

Maker's Marks: chalice w/o initials

Verification Stamps:

Adjustments: subtractions: slight routing on bottom of house

Form, Lid: split V-3 arm w/ handle

Form, House: straight; high type

Decoration, Lid: simple split V, squared; handle w/horses' heads

Decoration, House: geometric configurations

Dimensions: 6.4 x 4.4 x 4.9

Condition: incomplete; 1 missing cup

History: MRC/1973

Comments: This set was later labeled, inside the house and cups, in script which has corroded the metal slightly, to indicate the mass of each cup in grams: (490), (243), (123), (61), (30), (15), (7 1/2), (3), (1 1/2), respectively. Cf. Lockner, no. 70.

Catalog No.: 12

Collection No.: CIN6

Usage Type: Commercial

Century: 18th

Place of Manufacture: Nuremberg

Place of Use: France

Mass Quantity: 16 demi-once (16)/1/2 livre

Theoretical Mass: 489.5g per French pound

Numerical System: binary

Numerical Breakdown: 16, (8), (4), 2, 1 demi-once; 2, [1], [1/2], [1/2] gros

Actual Mass: 122.5g, 61.5g, 30.6g, 15.2g, 7.7g

Maker's Marks: halberd w/o initials

Verification Stamps:

Adjustments: none

Form, Lid: split V-3 arm w/o handle

Form, House: straight; low type

Decoration, Lid: plain

Decoration, House: plain

Dimensions: 4.3 x 2.8 x 2.6

Condition: incomplete; 3 missing cups

History: ECS/1940

Comments: Two cups appear extrinsic because of different patinas. They have the same verification stamps, however. It is possible that they were extrinsic to the set before being later verified. Ref. Kisch, 1965, p. 179; cf. Houben, 1984, p. 56, fig. 113.

Catalog No.: 13

Collection No.: CIN12

Usage Type: Commercial

Century: 17th

Place of Manufacture: Nuremberg

Place of Use: France

Mass Quantity: 1/4 livre

Theoretical Mass: 489.5g per French pound

Numerical System: unknown

Numerical Breakdown: unknown

Actual Mass: 60.8g

Maker's Marks: Paschal Lamb w/initials S.K.

Verification Stamps: one illegible mark and

Adjustments: none

Form, Lid: split V 3-arm w/o handle

Form, House: straight; low type

Decoration, Lid: simple split V, squared

Decoration, House: plain

Dimensions: 3.3 x 2.3 x 1.9

Condition: incomplete; house only

History: ECS/before 1940

Comments: There is an (8) stamped into the bottom of the house. Ref. Kisch, 1965, p. 182; cf. Houben, 1984, p. 59; cf. Schenk-Behrens, 1984, no. 550; cf. Lockner, no. 818.

Catalog No.: 14

Collection No.: CIN57

Usage Type: Commercial

Century: 18th

Place of Manufacture: Nuremberg

Place of Use: France

Mass Quantity: 1/4 livre

Theoretical Mass: 489.5g per French pound

Numerical System: binary

Numerical Breakdown: 4, 2, 1 once; 4, 2, 1, 1/2, [1/4], [1/4] gros

Actual Mass: 61.4g, 30.7g, 15.5g, 7.6g, 3.8g, 1.8g

Maker's Marks: none

Verification Stamps: none

Adjustments: none

Form, Lid: split V 3-arm w/o handle

Form, House: straight; low type

Decoration, Lid: simple split V, squared

Decoration, House: plain

Dimensions: 3.5 x 2.2 x 1.9

Condition: incomplete; 2 cups missing

History: MRC/1973

Comments:

Catalog No.: 15

Collection No.: CIN3

Usage Type: Bullion

Century: 18th

Place of Manufacture: Nuremberg

Place of Use: France

Mass Quantity: 16 livre (16)/32 marc

Theoretical Mass: 489.5g per French pound/244.75g per French marc

Numerical System: binary

Numerical Breakdown: 16, (8), 4, (2), (1) livre; 8 (oooo oooo) once/32, 16, 8 (VIII M), 4 (IIII M), 2 (II M), 1 marc; 4 (IIII O), 2 (II O), 1 (I O) once; [4], [2], [1], [1/2], [1/4], [1/4] gros

Actual Mass: 3920.0g, 1960.0g, 979.9g, 489.5g, 244.7g, 122.35g, 61.2g, 30.4g,

Maker's Marks: three crowns w/o initials

Verification Stamps:

Adjustments: subtractions: routing in 4, 2, 1 loth cups; additions: French coin weight is soldered onto lid between split V

Form, Lid: split V-3 arm w/ handle

Form, House: straight; low type

Decoration, Lid: handle w/ crown and horses' heads; knobs w/mermaids; hinge & lock w/seadogs

Decoration, House: geometric and pictorial configurations (hunters heads—2 types—and leaves)

Dimensions: 13.2 x 9.3 x 9.7

Condition: incomplete; 6 missing cups

History: ECS/1940

Comments: Ref. Kisch, 1965, p. 128, fig. 86, & p. 179; ref. Houben, 1984, p. 53, fig. 99; cf. Lavagne and Rochesnard, p. 84–85; cf. Lockner, no. 85.

Catalog No.: 16

Collection No.: CIN26

Usage Type: Bullion/Commercial

Century: 18th

Place of Manufacture: Nuremberg

Place of Use: France

Mass Quantity: 16 livre (16)/32 marc

Theoretical Mass: 489.5g per French pound/244.75g per French marc

Numerical System: binary

Numerical Breakdown: 16, (8), (4), (2), (1) livre; 8 once/32, 16, 8 (VIII MARS), 4 (IIII MARS), 2 (II MARS), 1 (I MAR) marc; 4 (IIII O), 2 (II O), 1 (I O) once; 4 (****), 2 (**), 1 (*), 1/2, [1/4], [1/4] gros

Actual Mass: 3917.8g, 1960.05g, 979.1g, 489.5g, 244.6g, 122.5g, 61.0g, 30.5g, 15.2g, 7.6g, 3.8g, 1.9g

Maker's Marks: key & arrow w/initials G.M (?)

Verification Stamps:

 T и

Adjustments: subtractions: routing; additions: one hole outside house is re-filled with brass

Form, Lid: split V-3 arm w/ handle

Form, House: waisted; high type

Decoration, Lid: handle w/ crown and horses' heads; knobs w/mermaids; split V w/ seadog-seahorse combination; hinge & lock w/seadog-seahorse combination

Decoration, House: roped; pictorial configurations (hunting scenes)

Dimensions: 12.9 x 9.0 x 10.6

Condition: incomplete; 2 missing cups

History: ECS/before 1940

Comments: Ref. Kisch, 1965, pp. 180 & 183; cf. Lavagne and Rochesnard, p. 84–85. cf. Lockner, no. 864.

Catalog No.: 17

Collection No.: CIN10

Usage Type: Commercial (Bullion?)

Century: 18th

Place of Manufacture: Nuremberg

Place of Use: France

Mass Quantity: 16 livre (16)

Theoretical Mass: 489.5g per French pound

Numerical System: unknown, probably binary

Numerical Breakdown: unknown; probably 16, (8), [4], [2], [1] livre . . .

Actual Mass: 3917.7g

Maker's Marks: deer (?) w/o initials

Verification Stamps: none

Adjustments: additions: 2 pieces of brass soldered on top of lid; one is a French coin weight

Form, Lid: split V-3 arm w/ handle

Form, House: straight; low type

Decoration, Lid: split V w/ seadogs; hinge & lock w/ seadogs; handle w/horses' heads; knobs w/bearded men

Decoration, House: geometric configurations

Dimensions: 12.8 x 9.2 x 9.5

Condition: incomplete; house only; lock is broken

History: ECS/1941–3; New York City purchase

Comments: Ref. Kisch, 1965, p. 181; ref. Houben, 1984, p. 28, fig. 44; cf. Lavagne and Rochesnard, p. 84–85; cf. Lockner, no. 58.

Catalog No.: 18

Collection No.: CIN9

Usage Type: Bullion/Commercial

Century: early 18th

Place of Manufacture: Nuremberg

Place of Use: France

Mass Quantity: 4 livre (4) /8 marc

Theoretical Mass: 489.5g per French pound/244.75g per French marc

Numerical System: binary

Numerical Breakdown: 4, (2), (1) livre; (16), {(8)}, [4], [2], (1) demi-once/8, 4 (IIII M), 2 (II M), 1 (I M) marc; {4 (oooo)}, [2], [1] once; (1), [1/2], [1/4], [1/4] gros

Actual Mass: 976.3g, 489.9g, 244.9g, {122.7g}, 15.6g,

Maker's Marks: lock w/o initials

Verification Stamps:

Adjustments: none

Form, Lid: split V-3 arm w/ handle

Form, House: slightly waisted; high type

Decoration, Lid: handle w/horses' heads; simple split V, squared

Decoration, House: roped; geometric configurations

Dimensions: 8.4 x 5.6 x 6.2

Condition: incomplete; 6 missing cups; 1 extrinsic cup

History: ECS/before 1940

Comments: Ref. Kisch, 1965, p. 128, fig. 86 & p. 178; cf. Houben, 1984, p. 59, fig. 130; cf. Lockner, no. 154.

Catalog No.: 19

Collection No.: CIN8

Usage Type: Bullion/Commercial

Century: early 18th

Place of Manufacture: Nuremberg

Place of Use: France

Mass Quantity: 2 livre (2)/4 marc

Theoretical Mass: 489.5g per French pound/244.75g French marc

Numerical System: binary

Numerical Breakdown: 2, (1) livre; (16), (8), (4), (2), (1) demi-once/4, 2, 1 (1 MARC) marc; 4 (IIII O), 2 (II O), 1 (I O) once; 4 (****), 2 (**), 1 (*), [1/2], [1/4], [1/4] gros

Actual Mass: 488.9g, 245.0g, 122.7g, 61.4g, 30.7g, 15.5g, 7.5g, 3.6g

Maker's Marks: scales w/o initials

Verification Stamps:

Adjustments: subtractions: routing outside house and cups

Form, Lid: split V-3 arm w/ handle

Form, House: straight; high type

Decoration, Lid: handle w/ horses' heads; simple split V, squared

Decoration, House: geometric configurations

Dimensions: 6.8 x 4.0 x 4.8

Condition: incomplete; 3 missing cups

History: ECS/before 1940

Comments: A handle on a nested weight of less than 4 pounds is unusual. Ref. Kisch, 1965, p. 179; cf. Houben, 1984, p. 59, fig. 133; cf. Lockner, no. 183.

Catalog No.: 20

Collection No.: BF15

Usage Type: Coin (from a coin scale and weight box)

Century: 18th (ca. 1775)

Place of Manufacture: Nuremberg

Place of Use: France

Mass Quantity: 8 demi-once (8) /4 once (4 ONCE)

Theoretical Mass: 489.5g per French pound

Numerical System: binary

Numerical Breakdown: 8, (4), (2), (1) demi-once/4, 2, 1 (I O) once; 4 (****), 2 (**), 1 (*), 1/2, 1/2 gros

Actual Mass: 122.2g, 61.2g, 30.6g, 15.3g, 7.6g, 3.8g, 1.9g, 1.9g

Maker's Marks: chalice w/o initials

Verification Stamps:

Adjustments: subtractions: filing

Form, Lid: split V-3 arm w/o handle

Form, House: straight; low type

Decoration, Lid: plain

Decoration, House: plain

Dimensions: 3.45 x 2.5 x 1.95

Condition: complete

History: ECS/1943–44

Comments: Handwritten in script in ink on the outside of house and on bottom of cups is the mass of each, mostly illegible, but roughly corresponding to the actual mass as given. The total mass of the whole set is 122.5g. Cf. Lockner, no.70.

Catalog No.: 21

Collection No.: CIN60

Usage Type: Coin (from a coin scale and weight box)

Century: 18th

Place of Manufacture: Nuremberg

Place of Use: Germany

Mass Quantity: 32 French crowns (32)

Theoretical Mass: 3.36g per French Crown

Numerical System: binary

Numerical Breakdown: 32, (16), (8), (4), (2), (1), 1/2, 1/4, 1/8, 1/16, 1/16 crown

Actual Mass: 108.2g, 53.3g, 27.9g, 13.2g, 6.6g, 3.4g, 1.3g, 0.7g, 0.35g, 0.175g, 0.175g

Maker's Marks: mermaid w/o initials

Verification Stamps:

17 67
K

Adjustments: none

Form, Lid: split V-3 arm w/o handle

Form, House: straight; low type

Decoration, Lid: simple split V, squared

Decoration, House: geometric configurations

Dimensions: 3.5 x 2.4 x 1.9

Condition: complete

History: ECS/1944

Comments: Catalog Nos. 21 and 22 are from the same coin scale and weight box. Cf. Lockner, no. 115.

Catalog No.: 22

Collection No.: CIN59

Usage Type: Coin (from a coin scale and weight box)

Century: 18th

Place of Manufacture: Nuremberg

Place of Use: Germany

Mass Quantity: 32 Hungarian gold ducats (32)

Theoretical Mass: 3.49g per Hungarian gold ducat

Numerical System: binary

Numerical Breakdown: 32, (16), (8), (4), (2), (1), 1/2, 1/4, 1/8, 1/16, 1/16 ducat

Actual Mass: 110.1g, 55.7g, 26.7g, 13.8g, 6.7g, 3.4g, 1.3g, 0.7g, 0.35g, 0.175g, 0.175g

Maker's Marks: mermaid w/o initials

Verification Stamps:

Adjustments: subtractions: filing

Form, Lid: split V-3 arm w/o handle

Form, House: straight; low type

Decoration, Lid: simple split V, squared

Decoration, House: geometric configurations

Dimensions: 3.5 x 2.4 x 1.9

Condition: complete

History: ECS/1944

Comments: Catalog Nos. 22 and 21 are from the same coin scale and weight box. The total mass of the whole set is 110.1g. Cf. Schenk-Behrens, 1985, no. 460; cf. Lockner, no. 115.

Catalog No.: 23

Collection No.: CIN16

Usage Type: Coin

Century: 18th

Place of Manufacture: Nuremberg

Place of Use: Germany

Mass Quantity: 128 Hungarian gold ducats (128)

Theoretical Mass: 3.49g per Hungarian gold ducat

Numerical System: binary

Numerical Breakdown: 128, (64), (32), (16), (8), (4), (2), (1), 1/2, [1/2] ducat

Actual Mass: 213.3g, 112.0g, 56.1g, 28.1g, 14.2g, 7.2g, 3.5g, 1.8g

Maker's Marks: chalice w/o initials

Verification Stamps:

85
17 M

Adjustments: subtractions: on the outside of the house, near the base, but not on the underside, is a wedge-shaped piece of brass that has been "sliced" off.

Form, Lid: split V-3 arm w/ handle

Form, House: waisted; low type

Decoration, Lid: simple split V, rounded

Decoration, House: geometric configurations

Dimensions: 5.0 x 3.3 x 2.6

Condition: incomplete; 1 missing cup; handle is missing

History: ECS/before 1940

Comments: Ref. Kisch, 1965, p. 181; cf. Lockner, no. 70.

Catalog No.: 24

Collection No.: CIN11

Usage Type: Coin (from a coin scale and weight box)

Century: 17th

Place of Manufacture: Nuremberg

Place of Use: Germany

Mass Quantity: 32 Hungarian gold ducats (32)

Theoretical Mass: 3.49g per Hungarian gold ducat

Numerical System: binary

Numerical Breakdown: 32, (16), (8), (4), (2), (1), 1/2, [1/4], [1/8], [1/16], [1/16] ducat

Actual Mass: 55.8g, 27.9g, 13.9g, 6.9g, 3.5g, 1.8g

Maker's Marks: running wolf w/o initials

Verification Stamps: none

Adjustments: none

Form, Lid: split V-3 arm w/o handle

Form, House: straight; low type

Decoration, Lid: simple split V, rounded

Decoration, House: geometric configurations

Dimensions: 3.2 x 2.1 x 2.1

Condition: incomplete; 4 missing cups

History: ECS/before 1940

Comments: Ref. Kisch, 1965, p. 181; cf. Houben, 1984, p. 64, fig. 152; cf. Schenk-Behrens, 1985, no. 460; cf. Lockner, no. 825.

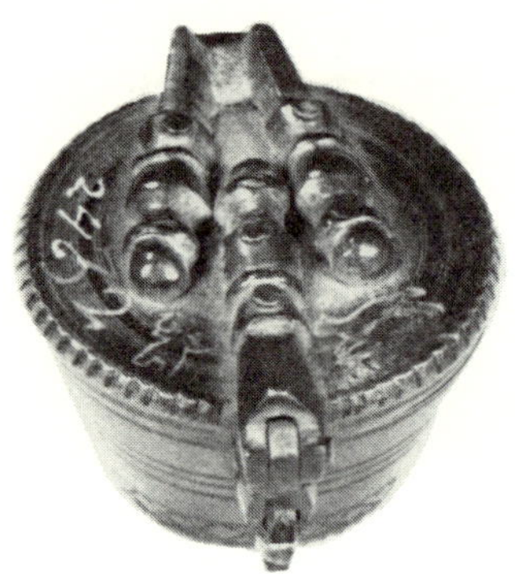

Catalog No.: 25

Collection No.: CIN23

Usage Type: Bullion

Century: 17th-18th

Place of Manufacture: Nuremberg

Place of Use: Germany

Mass Quantity: 32 pfund/64 marks (64)

Theoretical Mass: 472.4g per "light" Augsberg pound, or 490.87g per "heavy" Augsberg pound/235.9g per Mark

Numerical System: unknown

Numerical Breakdown: unknown

Actual Mass: approx. 7002.7g

Maker's Marks: split-tail mermaid w/initials I.A.S.

Verification Stamps:

A:M: 1799

AVGSPVRGER

1780

GWICHT

Adjustments: subtractions: handle, heads of mermaids, and part of the split V were removed from the lid

Form, Lid: split V-3 arm w/ handle

Form, House: slightly waisted; low type

Decoration, Lid: seadogs, mermaids

Decoration, House: roped; geometric and pictorial configurations (hunters, flying birds, trees, dogs, deer, boar)

Dimensions: 15.4 x 12.4 x 12.4

Condition: incomplete; house only; broken lock

History: unknown

Comments: This set was derived from the Augsberg standard but it is not the standard itself. Ref. Kisch, 1965, p. 183; cf. Conservatoire National des Arts et Métiers Catalogue, p. 106–7 and fig. 16; cf. Houben, 1984, p. 58, fig. 127; cf. Lockner, no. 1079.

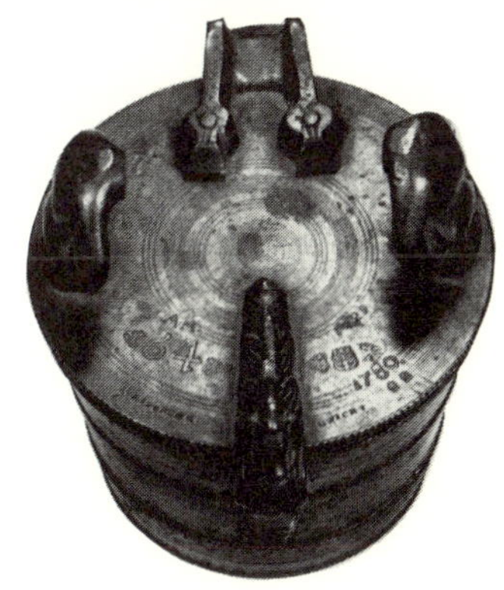

Catalog No.: 26

Collection No.: CIN15

Usage Type: Commercial (?)

Century: 18th

Place of Manufacture: Nuremberg

Place of Use: Germany (?)

Mass Quantity: 16 pounds (16)

Theoretical Mass: 475.0g per pound (?)

Numerical System: unknown

Numerical Breakdown: unknown

Actual Mass: 3805.7g

Maker's Marks: chalice w/o initials

Verification Stamps: flanking the chalice and the '16'

Adjustments: subtractions: routing and partial filling of the holes with lead

Form, Lid: split V-3 arm w/ handle

Form, House: slightly waisted; low type

Decoration, Lid: handle w/ horses' heads; split V w/ bearded, split-tail mermen; hinge w/seadogs; plain locker bar

Decoration, House: pictorial configurations (hunting scenes, split-tail mermaids, floral designs) and geometric configurations

Dimensions: 12.6 x 9.7 x 9.9

Condition: incomplete; house only; replaced locker bar; part of the hinge is missing

History: ECS/before 1940

Comments: According to Doursther, the Prussian pound is 475.6g and the Nuremberg mark is 238.6g (477.1g per pound). Therefore, it is possible that this set was used for bullion. At the base, under the hinge, the square stamp is repeated twice. Cf. Lockner, no. 70.

Catalog No.: 27

Collection No.: CIN52

Usage Type: Commercial

Century: 18th

Place of Manufacture: Nuremberg

Place of Use: Germany (N)

Mass Quantity: 4 pfund (4)

Theoretical Mass: 467.8g per Wurtemberg pound

Numerical System: binary

Numerical Breakdown: 4, (2), (1) pfund; (16), 8, (4), (2), (1), 1/2, {1/4}, {1/8}, [1/16], 1/16 loth

Actual Mass: 935.4g, 468.8g, 234.9g, 117.4g, 58.9g, 29.3g, 14.3g, 7.5g, {3.8g}, {2.1g}, 1.05g

Maker's Marks: stork w/ initials G.A.

Verification Stamps:

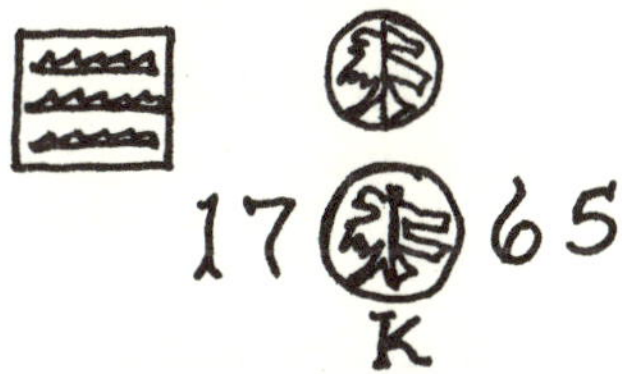

Adjustments: additions: lead soldered onto 16, 8, 4 loth cups and partial removal of same

Form, Lid: split V-3 arm w/ handle

Form, House: slightly waisted; high type

Decoration, Lid: handle w/ crown and horses' heads; simple split V, squared

Decoration, House: roped; geometric configurations

Dimensions: 8.5 x 5.4 x 6.1

Condition: incomplete; 1 missing cup; 2 extrinsic cups

History: MRC/1973

Comments: Although it is rare to see the innermost disk still intact with the set, this one appears more like the set than the two extrinsic cups before it. Cf. Lockner, no. 1499.

Catalog No.: 28

Collection No.: CIN22

Usage Type: Commercial

Century: 18th

Place of Manufacture: Nuremberg

Place of Use: Germany

Mass Quantity: 4 pfund (4)

Theoretical Mass: 467.8g per Wurtemberg pound

Numerical System: binary

Numerical Breakdown: 4, (2), (1) pfund; (16), (8), (4), (2), (1), [1/2], [1/4], [1/8], [1/16], [1/16] loth

Actual Mass: 914.6g, 467.3g, 233.5g, 116.5g, 58.1g, 28.9g, 14.4g

Maker's Marks: split-tail mermaid w/initials G.S.

Verification Stamps:

Adjustments: none

Form, Lid: split V-3 arm w/ handle

Form, House: slightly waisted; high type

Decoration, Lid: handle w/ horses' heads; simple split V, squared

Decoration, House: roped; geometric configurations

Dimensions: 8.2 x 5.4 x 6.7

Condition: incomplete; 5 missing cups; replaced handle

History: ECS/1943

Comments: The numerical verification stamps are on the inside bottom of the house. ref. Kisch, 1965, p. 183; cf. Lockner, no. 1219.

Catalog No.: 29

Collection No.: CIN55

Usage Type: Commercial

Century: early 19th

Place of Manufacture: Nuremberg

Place of Use: Bavaria (B)

Mass Quantity: 2 pfund (2)

Theoretical Mass: 560.06g per Bavarian pound

Numerical System: binary

Numerical Breakdown: 2, (1) pfund; (16), (8), (4), (2), (1), 1/2, 1/4, 1/8, 1/16, [1/16] loth

Actual Mass: 559.2g, 281.1g, 140.5g, 70.0g, 35.0g, 17.5g, 8.6g, 4.0g, 2.0g, 1.0g

Maker's Marks: chalice w/o initials

Verification Stamps:

Adjustments: subtractions: slight routing on bottom of house

Form, Lid: split V-3 arm w/o handle

Form, House: straight; high type

Decoration, Lid: simple split V, squared

Decoration, House: plain

Dimensions: 7.0 x 4.3 x 5.6

Condition: incomplete; 1 missing cup

History: MRC/1973

Comments: Cf. Lockner, no. 70; cf. Schenk-Behrens, 1987, no. 586.

Catalog No.: 30

Collection No.: CIN13

Usage Type: Bullion

Century: 18th

Place of Manufacture: Nuremberg

Place of Use: Hanover (H)

Mass Quantity: 16 loth (16)

Theoretical Mass: 486.7g per Hanover pound

Numerical System: binary

Numerical Breakdown: 16, (8), (4), (2), (1), 1/2, {1/4}, [1/8], [1/8] loth

Actual Mass: 118.1g, 59.2g, 29.9g, 15.0g, 7.3g, {3.8g}

Maker's Marks: seahorse w/o initials

Verification Stamps:

Adjustments: additions: brass added into 1 loth cup; subtractions: filing of metal from bottom of 2 loth cup

Form, Lid: split V-3 arm w/o handle

Form, House: straight; low type

Decoration, Lid: simple split-V, squared

Decoration, House: plain

Dimensions: 4.2 x 2.9 x 2.6

Condition: incomplete; 2 missing cups; 1 extrinsic cup

History: ECS/1947

Comments: The verification stamps have been put into the adjustment of the 1 loth cup. The lion stamp is also on the inside of the lid. Ref. Kisch, 1965, p. 181. Cf. Lockner, no. 116.

Catalog No.: 31

Collection No.: CIN20

Usage Type: Commercial

Century: 17th-18th

Place of Manufacture: Nuremberg

Place of Use: Germany

Mass Quantity: 16 loth (16 L)

Theoretical Mass: 467.7g per Cologne pound/233.89 per Cologne mark

Numerical System: binary

Numerical Breakdown: 16, (8), [4], [2], [1] loth; [2], [1], [1/2], [1/2] quent

Actual Mass: 115.5g

Maker's Marks: split-tail mermaid w/o initials

Verification stamps: none

Adjustments: none

Form, Lid: split V-3 arm w/o handle

Form, House: straight side; low type

Decoration, Lid: simple split V, squared

Decoration, House: plain

Dimensions: 3.9 x 2.3 x 2.5

Condition: incomplete; house only

History: ECS/1945–46

Comments: Cf. Schenk-Behrens, 1984, no. 549, and 1982, no. 552; cf. Lockner, no. 1306.

Catalog No.: 32

Collection No.: CIN2

Usage Type: Commercial

Century: 15th-16th (?)

Place of Manufacture: Nuremberg

Place of Use: Austria

Mass Quantity: 8 pfund (8)

Theoretical Mass: 560.0g per Viennese pound

Numerical System: binary

Numerical Breakdown: 8, (4), (2), (1) pfund; (16), (8), (4), (2), (1) loth; {2}, {1}, [1/2], {1/2} quenten

Actual Mass: 2249.0g, 1124.5g, 559.7g, 279.7g, 140.0g, 69.7g, 34.8g, 17.8g, {8.7g}, {4.2g}, {2.4g}

Maker's Marks: 3 crowns w/o initials (to the left of the locker)

Verification Stamps: none

Adjustments: subtraction: removal of split V; addition: brass piece soldered onto lid

Form, Lid: originally split V-3 arm w/handle

Form, House: straight; low type

Decoration, Lid: handle w/ crown; simple split-V, squared; replaced "arms" are unadorned

Decoration, House: plain; geometric configurations

Dimensions: 11.5 x 8.8 x 6.9

Condition: incomplete; 1 missing cup; 3 extrinsic cups; replaced "arm" on lid

History: ECS/1943–4; New York City purchase

Comments: Dr. Houben attributes this set to the 16th c.: 1525 marked the beginning of the so-called "remarkable pieces." Ref. Kisch, 1965, p. 179; cf. Houben, 1984, p. 53, fig. 99; cf. Lockner, no. 85.

Catalog No.: 33

Collection No.: CIN51

Usage Type: Commercial

Century: 18th-19th

Place of Manufacture: Nuremberg

Place of Use: Switzerland (?) or Austria (?)

Mass Quantity: 8 pfund (8)/ 16 Marks

Theoretical Mass: 537.8g per Thun (Swiss) pound of 18 ounces or 558.7g. per Fiume (Austro-Hungarian) pound/ 280.64g per Austrian or Tyrolian mark

Numerical System: binary

Numerical Breakdown: 8, 4, 2, 1 pound; 8 ounce/1 mark; 4 ounce /8 loth; 2 ounces/4 loth; {1} ounce/{2} loth; {1}, {1/2}, [1/4], [1/8], [1/8] half-ounce (see comments)

Actual Mass: 2235.7g, 1109.3g, 537.3g, 252.8g, 126.6g, 62.9g, {31.1g}, {15.6g}, {7.6g}

Maker's Marks: Paschal Lamb (three times) w/initials H. R.

Verification Stamps:

Adjustments: additions: lead soldered into cups and partial removal of same

Form, Lid: split V-3 arm w/ handle

Form, House: straight; high type

Decoration, Lid: handle w/ horses' heads; simple split V, squared

Decoration, House: roped; geometric configurations

Dimensions: 10.5 x 6.4 x 8.6

Condition: incomplete; 3 missing cups; 3 extrinsic cups

History: MRC/1973

Comments: Respectively, the cups are marked: (4), (XXXV), (I) and (XVII), (I M) and (VIII), (8) and (IIII), (4) and (II), {(2) and (I)}, {(1)}. There are no marks on {1/2} cup. According to Houben, 5–28–85, thirty-five may have been mistakenly used for thirty-three.

Catalog No.: 34

Collection No.: CIN17

Usage Type: Commercial

Century: 18th

Place of Manufacture: Nuremberg

Place of Use: Italy; Torino, or another city having frequent trade with France

Mass Quantity: 12 libbra (12)/24 marco

Theoretical Mass: 489.5g per French pound / 244.75g per French marc

Numerical System: duodecimal

Numerical Breakdown: 12, (6), (3), (2) libbra; (16), (8), (4), (2), {1}, {1/2}, [1/4], [1/8], [1/8] uncie/24, 12, 6 (6 MARCS), 4 (4 MARCS), 1 (MARC) marc; 4 (IIII O), 2 (II O), 1 (I O) once; {(4 GROS)}, {(2 GROS)}, [1], [1/2], [1/2] gros

Actual Mass: 2932.2g, 1466.1g, 980.0g, 244.2g, 122.8g, 61.2g, 30.7g, {14.8g}, {8.0g}

Maker's Marks: chalice w/o initials

Verification Stamps:

Adjustments: subtractions: routing on bottom of house and cups, lid deliberately damaged (handle and knob broken off)

Form, Lid: split V-3 arm w/handle

Form, House: slightly waisted; high type

Decoration, Lid: simple split V, rounded; simple hinge & lock

Decoration, House: roped; geometric configurations

Dimensions: 12.3 x 8.4 x 9.7

Condition: incomplete; 3 missing cups; 2 extrinsic cups; handle and one knob intentionally broken off for adjustment purposes

History: ECS/1944–6

Comments: According to Houben, 7–15–86, Torino probably ordered this set from Nuremberg in duodecimal for the pounds, but with a French mass in order to avoid disputes over the exchange rates. The set was later sent to Paris for verification. However, he says, it could also have been a standard set for Milan. Ref. Kisch, 1965, p. 178. cf. Lockner, no. 70.

Catalog No.: 35

Collection No.: CIN14

Usage Type: Commercial

Century: 18th

Place of Manufacture: Nuremberg

Place of Use: Northeastern Italy, Vienna (W), and Breslau (B)

Mass Quantity: 12 libbra

Theoretical Mass: 405.0g per Breslau pound

Numerical System: duodecimal and binary

Numerical Breakdown: 12, (6), (3), (2) libbra; (16), (8), (4), (2), (1), 1/2, 1/4, 1/8, 1/8 loth

Actual Mass: 4858.4g, 2429.2g, 1214.6g, 808.3g, 202.5g, 101.2g, 50.5g, 25.5g, 12.7g, 6.1g, 3.1g, 1.6g, 1.6g

Maker's Marks: seahorse w/o initials

Verification Stamps:

Adjustments: subtractions: filing on bottom of cups

Form, Lid: split V-3 arm w/ handle

Form, House: waisted; high type

Decoration, Lid: handle w/ horses' heads and crown; simple split V, rounded

Decoration, House: roped; geometric and pictorial configurations (running dogs)

Dimensions: 12.3 x 8.1 x 8.3

Condition: complete

History: ECS/1945, New York City purchase

Comments: According to Houben, 5–28–85, the double eagle as well as the single eagle could be for Poland and the lion for Prussia. Therefore, it could have been made for "Austrian-Italy" for trade with Breslau. Ref. Kisch, 1965, p. 181; cf. Houben, 1984, p. 60, fig. 135. cf. Lockner, no. 35.

Catalog No.: 36

Collection No.: CIN28

Usage Type: Commercial/ Bullion

Century: 17th-18th

Place of Manufacture: Nuremberg

Place of Use: Northwestern Italy

Mass Quantity: 128 oncie/16 marcos (8 M)

Theoretical Mass: 368.84g per Turin pound of 12 ounces/245.896g per Turin marc

Numerical System: binary

Numerical Breakdown: 128 (100.28.0.TOTALE), 64 (64 ONCIE PIEMO)/8 (8 M) marcos, 32 (32 ONCIE PIEMONTE), (16), 8 (8 oooooooo), (4), (2), (1) oncie; 4, 2, 1, [1/2], [1/2] ottavio

Actual Mass: 1965.2g, 982.5g, 492.7g, 246.1g, 123.1g, 61.5g, 30.5g, 15.4g, 7.7g, 3.8g

Maker's Marks: Key & arrow w/initials G. S. (?)

Verification Stamps:

Adjustments: additions: brass piece soldered in between split V, very small pieces of brass soldered into cups

Form, Lid: split v-3 arm w/ handle

Form, House: straight; high type

Decoration, Lid: handle w/ horses heads & crown; simple hinge & lock; simple split V,

Decoration, House: roped; geometric configurations

Dimensions: 10.0 x 6.9 x 7.8

Condition: incomplete; 2 missing cups

History: ECS/before 1940

Comments: Piemonte = Turin; northern towns in Italy often used the binary system instead of the duodecimal system. The inside of this set is covered with verification stamps. Ref. Kisch, 1965, p. 183; cf. Houben, 1984, p. 57, fig. 120; cf. Lockner, no. 715; cf. Lockner, no. 715.

Catalog No.: 37

Collection No.: CIN56

Usage Type: Commercial

Century: 18th

Place of Manufacture: Nuremberg

Place of Use: Italy; Venice (V) and Florence (?) (F)

Mass Quantity: 24 oncie (24)

Theoretical Mass: 338.9g per Padua or 339.55g per Florence pound of 12 ounces

Numerical System: duodecimal

Numerical Breakdown: 24, (12), (6), (3), (2), (1/2), 1/4, 1/8, [1/8] oncie

Actual Mass: 339.7, 169.7, 85.2, 56.7, 14.1, 7.3, 3.7, 2.1g

Maker's Marks: chalice w/o initials

Verification Stamps: none

Adjustments: subtractions: filing on bottom of cup and house

Form, Lid: split V-3 arm w/o handle

Form, House: slightly waisted; high type

Decoration, Lid: simple split V, squared

Decoration, House: geometric configurations

Dimensions: 4.9 x 3.8 x 4.3

Condition: incomplete; 1 missing cup

History: MRC/1973

Comments: The digit "4" is stamped into the bottom of the 6 oncie cup. Ref. Houben, 1984. p. 17, fig. 27; cf. Lockner, no. 70.

Catalog No.: 38

Collection No.: CITt2

Usage Type: Commercial

Century: 18th-19th (before 1870)

Place of Manufacture: Italy

Place of Use: Italy

Mass Quantity: 2 libbra

Theoretical Mass: 320.0g or more per Italian pound of 12 ounces (See comments)

Numerical System: duodecimal

Numerical Breakdown: 24, [12], 6 (oooooo), 3, 2 (oo), {1/2}, {1/4}, [1/8], [1/8] oncie

Actual Mass: 159.3g, 78.7g, 51.3g, {12.3g}, {6.3g}

Maker's Marks: none

Verification Stamps: none

Adjustments: subtractions: routing on bottom of 1/2 oncie cup

Form, Lid: unknown

Form, House: straight

Decoration, Lid: unknown

Decoration, House: plain

Dimensions: 5.9 x 4.2 x 3.9

Condition: incomplete; 2 missing cups; 2 extrinsic cups; lid of the house is missing; the set is quite worn

History: ECS/before 1940

Comments: The house—without the lid, which is missing—weighs 354.9g, suggesting a higher theoretical mass than twice the 6 oz. weight (318.6g) would indicate. Cf. Houben 1984, p. 31, fig. 50.

Catalog No.: 39

Collection No.: CITt3

Usage Type: Commercial

Century: 18th-19th (before 1870)

Place of Manufacture: Italy

Place of Use: Italy

Mass Quantity: 1 libbra

Theoretical Mass: at least 320.0g per Italian pound of 12 ounces (see comments)

Numerical System: duodecimal

Numerical Breakdown: 12, 6 (oooooo), 3 (ooo), 2 (oo), 1/2, 1/4, 1/8, [1/8] oncie

Actual Mass: 158.5g, 79.1g, 53.1g, 13.2g, 6.7g, 3.3g

Maker's Marks: none

Verification Stamps:

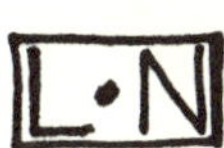

Adjustments: additions: lead soldered into cups and house

Form, Lid: no handle, flattened cone with prominent knob

Form, House: straight; high type

Decoration, Lid: plain (concentric rings in deep relief)

Decoration, House: plain

Dimensions: 4.5 x 3.3 x 3.9

Condition: incomplete; 1 missing cup

History: MRC/1973

Comments: The Milan "light" pound is 326.8g and the Naples pound is 320.76g. Cf. Houben, 1984, p. 31, fig. 50.

Catalog No.: 40

Collection No.: CITt4

Usage Type: Commercial

Century: 18th-19th (before 1879)

Place of Manufacture: Italy

Place of Use: Italy

Mass Quantity: 1 libbra

Theoretical Mass: at least 320.og per Italian pound of 12 ounces (See comments)

Numerical System: duodecimal

Numerical Breakdown: 12, 6 (oooooo), 3 (ooo), {2 (oo)}, {1/2}, [1/4], [1/8], [1/8] oncie

Actual Mass: 158.6g, 79.2g, {52.0g}, {14.0g}

Maker's Marks: none

Verification Stamps: none

Adjustments: additions: metal soldered into 1/2 oncie cup

Form, Lid: no handle, flattened cone

Form, House: straight; high type

Decoration, Lid: plain

Decoration, House: plain

Dimensions: 4.4 x 2.6 x 3.7

Condition: incomplete; 3 missing cups; 2 extrinsic cups

History: ECS/before 1940

Comments: The Milan "light" pound is 326.8g and the Naples pound is 320.76g. Cf. Houben, 1984, p. 31, fig. 50.

Catalog No.: 41

Collection No.: CITt5

Usage Type: Commercial

Century: 18th-19th (before 1870)

Place of Manufacture: Italy; Naples

Place of Use: Italy

Mass Quantity: 1 libbra

Theoretical Mass: 320.6g per Naples pound

Numerical System: duodecimal

Numerical Breakdown: 12, 6, 3, 2, 1/2, 1/4, {1/8}, [1/8] oncie

Actual Mass: 160.4g, 80.5g, 53.6g, 13.6g, 6.9g, {7.0g}

Maker's Marks: none

Verification Stamps: none

Adjustments: none

Form, Lid: no handle, flattened cone with prominent knob

Form, House: straight; high type

Decoration, Lid: plain

Decoration, House: plain

Dimensions: 4.2 x 2.8 x 4.2

Condition: incomplete; 1 missing cup; disk is extrinsic to the set

History: ECS/before 1940

Comments: Cf. Houben, 1984, p. 31, fig. 50.

Catalog No.: 42

Collection No.: CIF3

Usage Type: Commercial

Century: 18th

Place of Manufacture: Italy

Place of Use: Italy

Mass Quantity: 1 libbra

Theoretical Mass: 318.0g per Plaisance or Piacenza pound

Numerical System: duo-decimal

Numerical Breakdown: 30, 12, 6, 3, 2, 1/2, 1/4, 1/8, 1/8 oncie

Actual Mass: Lid = 158.5g, House = 316.8g, 316.8g, 158.8g, 79.1g, 52.5g, 12.8g, 6.9g, 3.1g, 3.1g

Maker's Marks: none

Verification Stamps:

Adjustments: none

Form, Lid: screw on lid with prominent knob

Form, House: straight; high type

Decoration, Lid: plain

Decoration, House: plain

Dimensions: 6.2 x 3.8 x 5.5

Condition: complete

History: ECS/before 1940

Comments: Cf. Houben, 1984, p. 31, fig. 49.

Catalog No.: 43

Collection No.: CITt1

Usage Type: Commercial

Century: 18th (before 1870)

Place of Manufacture: Italy

Place of Use: Italy

Mass Quantity: 1 libbra

Theoretical Mass: 330.0g per Reggio pound (see comments)

Numerical System: duodecimal

Numerical Breakdown: 12, 6, {3}, {2}, {1/2}, [1/4], {1/8}, {1/8} oncie

Actual Mass: 165.1g, {79.9g}, {53.5g}, {11.5g}, {3.3g}, {3.2g}

Maker's Marks: none

Verification Stamps: none

Adjustments: none

Form, Lid: no handle, flattened cone with prominent knob

Form, House: straight; high type

Decoration, Lid: plain

Decoration, House: plain

Dimensions: 4.2 x 2.3 x 4.5

Condition: incomplete; 1 missing cup; 5 extrinsic cups; lid does not close because cups do not fit properly

History: ECS/before 1940

Comments: Ancona and St. Remo are 331.0g and Verona is 333.0g. The cities in Italy have such various but similar theoretical masses that it is impossible to give an accurate provenance. Cf. Houben, 1984, p. 31, fig. 49.

Catalog No.: 44

Collection No.: CIA1

Usage Type: unknown

Century: 19th

Place of Manufacture: Austria

Place of Use: Austria

Mass Quantity: unknown

Theoretical Mass: 560.1g per Viennese pound

Numerical System: unknown

Numerical Breakdown: 1, 1/2 loth

Actual Mass: 17.5g, 8.6g

Maker's Marks: none

Verification Stamps:

Adjustments: subtractions: filing on bottom of cups

Form, Lid: unknown

Form, House: unknown

Decoration, Lid: unknown

Decoration, House: unknown

Dimensions: 2.7 x 1.8 x 1.3

Condition: incomplete; house and all but 2 cups are missing

History: unknown

Comments: These two cups probably do not belong together. Cf. Schenk-Behrens, 1985, no. 459.

Catalog No.: 45

Collection No.: CIG3

Usage Type: Commercial

Century: 19th

Place of Manufacture: Germany

Place of Use: Germany (CROSSEN)

Mass Quantity: 1 pfund (1 PFUND)

Theoretical Mass: 500g per Prussian "Zollpfund"

Numerical System: pre-decimal

Numerical Breakdown: 1 pfund; 10 (10 L), 10 (10 L), 5 (5 L), 2 (2 L), 1 (1 L), 1 (1 L) loth; 5 (5 Q), 2 (2 Q), [2], [1] quint

Actual Mass: 499.1g, 165.7g, 166.8g, 83.2g, 33.2g, 16.7g, 16.7g, 8.3g, 3.2g

Maker's Marks: none

Verification Stamps:

1856 

Adjustments: none

Form, Lid: single-arm w/o handle; the "arms" broaden and there is a circle in between them

Form, House: straight; high type

Decoration, Lid: plain

Decoration, House: plain

Dimensions: 5.4 x 3.0 x 4.3

Condition: incomplete; 2 missing cups

History: MRC/1973

Comments: This "pre-decimal" transitional piece, dated 1856, was used by the Deutscher Zollverein. According to Batz, 11–11–86, Crossen was a town in Germany, "Krossen an der Oder" until 1945. The town is now Polish. Cf. Houben, 1984, p. 33, fig. 58; cf. Schenk-Behrens, 1987, no. 596.

Catalog No.: 46

Collection No.: CIG2

Usage Type: Commercial

Century: 19th (after 1872)

Place of Manufacture: Germany

Place of Use: Germany

Mass Quantity: 500 gramm

Theoretical Mass: 500g (500 G)

Numerical System: decimal

Numerical Breakdown: 500, (1/2 lb), (100 G), (50 G), (50 G), (20 G), (10 G), (10 G), (5 G), (2 G), (2 G), [1] grams

Actual Mass: 250.5g, 100.0g, 50.0g, 50.0g, 20.0g, 10.0g, 10.0g, 5.0g, 2.0g, 2.0g

Maker's Marks: none

Verification Stamps:

Adjustments: none

Form, Lid: single-arm w/o handle; the "arms" broaden and there is a circle in between them

Form, House: straight; low type

Decoration, Lid: plain

Decoration, House: plain

Dimensions: 5.6 x 3.5 x 3.9

Condition: incomplete; 1 missing cup

History: ECS/1943–4

Comments: According to Houben, 1984, p. 33, this is a later transitional piece, one of the first metric nested weights made in Germany.

Catalog No.: 47

Collection No.: CIG1

Usage Type: Commercial

Century: 19th (after 1880)

Place of Manufacture: Germany

Place of Use: Germany

Mass Quantity: 500 gramm (500 G)

Theoretical Mass: 500g

Numerical System: decimal

Numerical Breakdown: 500, (200 G), (100 G), (100 G), (50 G), (20 G), (10 G), (10 G), [5], [2], [1], [1] gram

Actual Mass: 200.0g, 100.0g, 100.0g, 50.0g, 20.0g, 10.0g, 10.0g

Maker's Marks: none

Verification Stamps:

Adjustments: none

Form, Lid: single-arm w/o handle

Form, House: straight; low type

Decoration, Lid: the "arms" are scalloped ("modified French metric type," according to Houben)

Decoration, House: plain

Dimensions: 5.6 x 3.1 x 3.9

Condition: incomplete; 4 missing cups

History: ECS/before 1940

Comments: Cf. Houben, 1984, p. 33, fig. 59.

Catalog No.: 48

Collection No.: CIF1

Usage Type: Commercial

Century: 19th

Place of Manufacture: France

Place of Use: France

Mass Quantity: 2 kilogrammes (2 KILOG)

Theoretical Mass: 2000g

Numerical System: decimal

Numerical Breakdown: 2, 1 kilogram (1000 GRAM and 1 KILOGRAMME); 500 (500 GRAMMES), 200 (200 GRAM), 100 (100 GRAM), 100 (100 GRAM), 50 (50 GRAM), 20 (20 GR), 10 (10 GR), 10 (10 GR), 5 (5 GR), [2], [1], [1] gram

Actual Mass: 1000.0g, 500.0g, 200.0g, 100.0g, 50.0g, 20.0g, 10.0g, 10.0g, 5.0g

Maker's Marks: crowned "H"

Verification Stamps:

Adjustments: none

Form, Lid: single-arm w/o handle

Form, House: straight; high type

Decoration, Lid: the "arm" broadens into a solid diamond

Decoration, House: plain

Dimensions: 8.4 x 5.7 x 6.3

Condition: incomplete; 3 missing cups

History: ECS/before 1940

Comments: According to Lavagne (collection notes made in 1972) the weight maker was Hamelin, who worked in Paris from 1802 to 1845. The digit "1" is imprinted over incorrect digit "5" in "1000 GRAM" inside the lid. Cf. Schenk-Behrens, 1982, no. 558.

Catalog No.: 49

Collection No.: CIF5

Usage Type: Commercial

Century: 19th

Place of Manufacture: France

Place of Use: France

Mass Quantity: 500 grammes (500 GRAMMES)

Theoretical Mass: 500g

Numerical System: decimal

Numerical Breakdown: 500, (200 GRAMMES), (100 GRAMMES), (100 GRAMMES), (50 GRAMMES), (20 GRAMMES), {10}, {10}, {5}, {2}, [1], [1] grammes

Actual Mass: 196.2g, 97.4g, 97.5g, 48.2g, 18.5g, {15.3g}, {7.8g}, {3.8g}, {1.9g}

Maker's Marks: crowned "M"

Verification Stamps:

Adjustments: none

Form, Lid: single-arm w/o handle

Form, House: straight; low type

Decoration, Lid: the "arm" broadens into a diamond

Decoration, House: plain

Dimensions: 5.9 x 4.4 x 2.9

Condition: incomplete; 2 missing cups; 4 extrinsic cups

History: ECS/before 1940

Comments: The extrinsic cups are 4 (****), 2 (**), 1 (*), 1/2 gros, respectively. According to Lavagne (Collection notes made in 1972) the weight maker is Montier, who worked in Paris from 1831 to 1853. Ref. Houben, 1984, p. 36, fig. 68; cf. Schenk-Behrens, 1982, no. 558.

Catalog No.: 50

Collection No.: CIF2

Usage Type: Commercial

Century: 19th (1812–1840)

Place of Manufacture: France

Place of Use: France

Mass Quantity: 1 livre (1 LIVRE)

Theoretical Mass: 489.5g per French pound

Numerical System: binary

Numerical Breakdown: 1 livre; 8, 4 (4 ONCES), 2 (2 ONCE), 1 (1 ONCE) once; 4 (4 GROS), 2 (2 GROS), 1 (1 GROS), 1/2 (DEMI), 1/2 (DEMI) gros

Actual Mass: 489.7g, 244.7g, 122.5g, 60.9g, 30.4g, 15.3g, 7.6g, 3.8g, 1.9g, 1.9g

Maker's Marks: none

Verification Stamps: none

Adjustments: none

Form, Lid: single-arm w/o handle

Form, House: straight; low type

Decoration, Lid: the "arm" broadens into a rectangle

Decoration, House: plain

Dimensions: 5.4 x 3.7 x 3.5

Condition: complete

History: MRC/1973

Comments: This set appears identical to Catalog No. 51, of the Système Métrique Usuel. However, its total mass is only 489.7g which is closer to the pre-metric pound.

Catalog No.: 51

Collection No.: CIF4

Usage Type: Commercial

Century: 19th (1812–1840)

Place of Manufacture: France

Place of Use: France

Mass Quantity: 1 livre (1 LIVRE)/500 grammes (500 GRAMME)

Theoretical Mass: 500g per French decimal pound

Numerical System: Système Métrique Usuel

Numerical Breakdown: 1 livre; 8 (8 ONCES), 4 (4 ONCES), 2 (2 ONCES), 1 (1 ONCE) once; 4 (4 GROS), 1 (1 GROS), 2 (2 GROS), 1/2 (DEMI), 1/2 (DEMI) gros

Actual Mass: 499.8g, 249.5g (250 GRAMMES), 125.0g (125 GRAMMES), 62.5g (62 G 5), 31.3g (31 G 3), 15.6g (15 G 6), 7.8g (7 G 8), 3.9g (3 G 9), 1.96g (1 G 96), 1.96g (1 G 96)

Maker's Marks: crowned "R"

Verification Stamps:

Adjustments: none

Form, Lid: single-arm w/o handle

Form, House: straight; low type

Decoration, Lid: the "arm" broadens into a rectangle

Decoration, House: plain

Dimensions: 5.4 x 3.5 x 3.5

Condition: complete

History: ECS/1944–45

Comments: The ONCE marks are on the rims; the GRAMME marks are on the inside bottom of the cups. According to Lavagne's Collection notes made in 1972, the weight maker was Ribon, who worked in Paris from 1834 to 1851. Cf. Houben, 1984, p. 37, fig. 71; cf. Brown, no. 108.

Catalog No.: 52

Collection No.: BF3

Usage Type: Coin (from a coin scale and weight box)

Century: 18th

Place of Manufacture: Paris, France

Place of Use: France

Mass Quantity: 1 once

Theoretical Mass: 489.5g

Numerical System: binary

Numerical Breakdown: 1 once; 4(****), 2(**), 1(*), 1/2, 1/2 gros

Actual Mass: 30.55g, 15.3g, 7.5g, 3.7g, 1.8g, 1.8g

Maker's Marks: none

Verification Stamps: none

Adjustments: filing

Form, Lid: lidless

Form, House: open set

Decoration, Lid: none

Decoration, House: none

Dimensions: 2.6 x 2.0 x 1.0

Condition: complete

History: BZK/1965

Comments: This set of weights is not original to the scale and weight box; it does not fit tightly enough into it. Ref. Kisch, 1965, p.127, fig. 85; cf. Houben, 1982, p. 40, fig. 119.

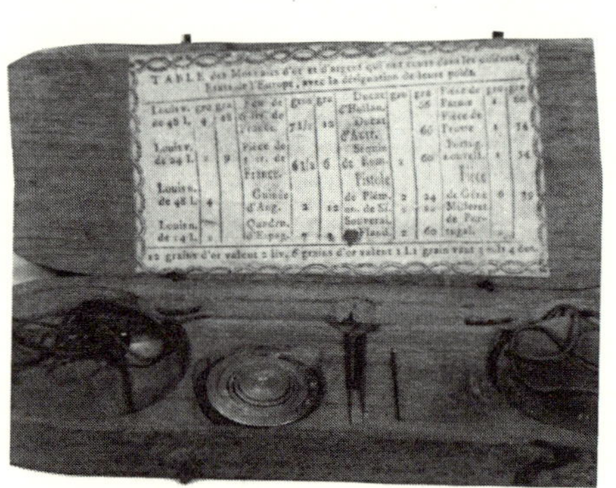

Catalog No.: 53

Collection No.: BF14

Usage Type: Coin (from a coin scale and weight box)

Century: 18th

Place of Manufacture: Paris, France

Place of Use: France

Mass Quantity: 1 once

Theoretical Mass: 489.5g

Numerical System: binary

Numerical Breakdown: 1 once; (4 GROS), (2 GROS), (1 GROS); (DEMI), (DEMI)

Actual Mass: 31.2g, 15.6g (15G6), 7.8g (7G8), 3.9g (3G9), 1.95g (1G95), 1.95g (1G95)

Maker's Marks: none

Verification Stamps:

Adjustments: filing

Form, Lid: lidless

Form, House: open set

Decoration, Lid: none

Decoration, House: none

Dimensions: 2.7 x 2.2 x 1.0

Condition: complete

History: ECS/before 1940

Comments: Cf. Houben, 1982, p. 40, fig. 119.

Catalog No.: 54

Collection No.: BF2

Usage Type: Coin (from a coin scale and weight box)

Century: 18th

Place of Manufacture: Paris, France

Place of Use: France

Mass Quantity: 1 once

Theoretical Mass: 489.5g per French pound

Numerical System: binary

Numerical Breakdown: 1 once; 4 (oooo), 2 (oo), 1 (o), 1/2, 1/2 gros

Actual Mass: 30.6g, 15.6g, 7.8g, 3.9g, 1.95g, 1.95g

Maker's Marks: none (see comments)

Verification Stamps: none

Adjustments: none

Form, Lid: lidless

Form, House: open set

Decoration, Lid: none

Decoration, House: none

Dimensions: 2.7 x 2.2 x 0.9

Condition: complete

History: ECS/1943

Comments: The label inside the lid of the box gives the maker's name as CHEMIN. Cf. Houben, 1982, p. 40, fig. 119.

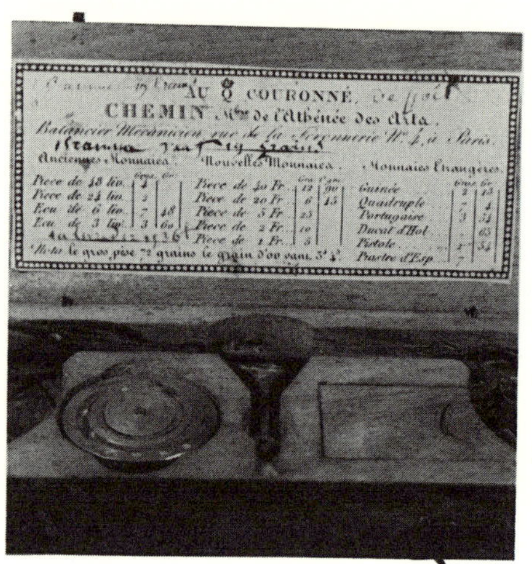

Catalog No.: 55

Collection No.: BF7

Usage Type: Coin (from a coin scale and weight box)

Century: late 18th–19th

Place of Manufacture: Bordeaux, France

Place of Use: France

Mass Quantity: 1 once

Theoretical Mass: 489.5g per French pound

Numerical System: binary

Numerical Breakdown: 1 once; 4 (*4*GROS*) and (2 LOUIS), 2 (*2*GROS*) and (1 LOUIS), [1], [1/2], [1/2] gros

Actual Mass: 15.6g, 7.8g

Maker's Marks: none (see comments)

Verification Stamps:

Adjustments: none

Form, Lid: lidless

Form, House: open set

Decoration, Lid: none

Decoration, House: none

Dimensions: 2.75 x 2.2 x 0.9

Condition: incomplete; 3 missing cups

History: BZK/1955

Comments: The label inside the lid gives the maker's name as A FOURNEL, from Bordeaux.

Catalog No.: 56

Collection No.: BF13

Usage Type: Coin (from a coin scale and weight box)

Century: 18th

Place of Manufacture: Paris, France

Place of Use: France

Mass Quantity: 1 once

Theoretical Mass: 489.5g per French pound

Numerical System: binary

Numerical Breakdown: 1 once; 4 (****), 2 (**), 1 (*), 1/2, 1/2 gros

Actual Mass: 31.2g, 15.6g, 7.8g, 3.9g, 1.95g, 1.95g

Maker's Marks: none

Verification Stamps: none

Adjustments: subtractions: filing

Form, Lid: lidless

Form, House: open set

Decoration, Lid: none

Decoration, House: none

Dimensions: 2.7 x 2.2 x 0.9

Condition: complete

History: ECS/1943

Comments: Cf. Houben 1982, p. 40, fig. 119.

Catalog No.: 57

Collection No.: BF10

Usage Type: Coin (from a coin scale and weight box)

Century: 18th

Place of Manufacture: Paris, France

Place of Use: France

Mass Quantity: 1 once

Theoretical Mass: 489.5g per French pound

Numerical System: binary

Numerical Breakdown: 1 once; 4 (****), 2 (**), 1 (*), 1/2, {1/2} gros

Actual Mass: 31.2g, 15.6g, 7.6g, 3.8g, 1.9g, {1.8g}

Maker's Marks: none

Verification Stamps: none

Adjustments: subtractions: filing

Form, Lid: lidless

Form, House: open set

Decoration, Lid: none

Decoration, House: none

Dimensions: 2.7 x 2.1 x 0.9

Condition: incomplete; 1 extrinsic cup

History: ECS/before 1940

Comments: Cf. Houben, 1982, p. 40, fig. 119.

Catalog No.: 58

Collection No.: CIA4

Usage Type: Commercial

Century: 19th

Place of Manufacture: Spain

Place of Use: Spain; Madrid

Mass Quantity: 4 libra

Theoretical Mass: 460.0g per Madrid pound

Numerical System: binary

Numerical Breakdown: 4, 2, 1 libra; 8, 4, 2, {1}, {1/2}, {1/4}, {1/8}, [1/8] onza/{4}, {2}, {1}, [1] ochava

Actual Mass: 919.8g, 460.5g, 230.7g, 114.7g, 57.4g, {28.7g}, {14.6g}, {7.3g}, {3.4g}

Maker's Marks: none

Verification Stamps:

 883

Adjustments: none

Form, Lid: single-arm w/o handle

Form, House: straight; high type

Decoration, Lid: the "arm" broadens into a split diamond ("lozenge-arm imitation")

Decoration, House: plain

Dimensions: 6.4 x 3.4 x 5.4

Condition: incomplete; 1 missing cup; 4 extrinsic cups

History: ECS/1945–46

Comments: According to Houben, 5–28–85, this set could have been later verified in Mexico. Ref. Houben, 1984, p. 35, fig. 66; cf. Lavagne, 1981, p. 154-55.

Catalog No.: 59

Collection No.: CIA3

Usage Type: Commercial

Century: 19th

Place of Manufacture: Spain

Place of Use: Spain; Madrid

Mass Quantity: 2 libra

Theoretical Mass: 460.0g per Madrid pound

Numerical System: binary

Numerical Breakdown: 2, 1 libra; 8, 4, [2], [1], [1/2], [1/4], [1/8], [1/8] onza/[4], [2], [1], [1] ochava

Actual Mass: 461.5g, 229.2g, 114.4g

Maker's Marks: none

Verification Stamps:

887 894

Adjustments: none

Form, Lid: single-arm w/o handle

Form, House: straight; high type

Decoration, Lid: the "arm" broadens into a split diamond ("lozenge-arm imitation")

Decoration, House: plain

Dimensions: 6.5 x 3.8 x 5.0

Condition: incomplete; 6 missing cups

History: CCF/1958

Comments: Cf. Lavagne, 1981, p. 154–55; cf. Houben, 1984, p. 35, fig. 66; cf. Schenk-Behrens, 1981, no. 403.

Catalog No.: 60

Collection No.: CIA5

Usage Type: Commercial

Century: 18th-19th

Place of Manufacture: Spain

Place of Use: Spain; Madrid

Mass Quantity: 1 libra

Theoretical Mass: 460.0g per Madrid pound

Numerical System: binary

Numerical Breakdown: 1 libra; 8, 4, {2}, 1 onza; 4, [2], [1], [1] ochava

Actual Mass: 231.0g, 113.5g, 55.9g, 28.7g, 14.1g

Maker's Marks: none

Verification Stamps:

886

P 889

895 897 P

891 887

888

Adjustments: none

Form, Lid: single-arm w/o handle ("three-arm imitation")

Form, House: straight; low type

Decoration, Lid: plain

Decoration, House: plain

Dimensions: 5.0 x 3.3 x 3.9

Condition: incomplete; 3 missing cups; 1 extrinsic cup; lid does not close flush with house.

History: ECS/1941–42

Comments: Some of the verification stamps are on the outside walls of the cups. The inside bottoms are crowded with them. According to Houben, 5–28–85, the set could have been later verified in Mexico. Cf. Houben, p. 35, fig. 65; cf. Lavagne, 1981, p. 154–55.

Catalog No.: 61

Collection No.: CIGB12

Usage Type: Commercial

Century: 18th-19th

Place of Manufacture: Spain

Place of Use: Spain

Mass Quantity: 1 libra

Theoretical Mass: 460.0g per Spanish pound (see comments)

Numerical System: binary

Numerical Breakdown: 1 libra; 8 (8 OZ), 4 (4 OZ), 2 (2 OZ), {1 (1 OZ)} onza; {4}, {2}, {1}, [1] ochava/{1/2}, {1/4}, {1/8}, [1/8] onza

Actual Mass: 230.0g, 115.2g, 57.5g, {30.8g}, {15.3g}, {7.9g}, {3.8g}

Maker's Marks: none

Verification Stamps: none

Adjustments: none

Form, Lid: single-arm w/o handle ("three-arm imitation")

Form, House: straight; low type

Decoration, Lid: plain

Decoration, House: plain

Dimensions: 5.1 x 3.6 x 3.8

Condition: incomplete; 1 missing cup; 4 extrinsic cups; lid does not close flush with house

History: ECS/1944–45

Comments: Madrid is 460.0g and Castile is 460.5g. Ref. Houben, 1984, p. 35, fig. 65.

Catalog No.: 62

Collection No.: CISp1

Usage Type: Commercial

Century: 19th

Place of Manufacture: Spain

Place of Use: Spain

Mass Quantity: 1 libra

Theoretical Mass: 460.0g per Spanish pound (See comments)

Numerical System: binary

Numerical Breakdown: 1 libra; [8], 4 (IIII ONZA), 2 (2 ONZA), 1 (1 ONZA) onza; 4 (IIII OCH), 2 (II OCH), 1 (I OCH), 1/2, 1/4, [1/8], [1/8] ochava

Actual Mass: 115.5g, 57.5g, 28.6g, 14.3g, 7.2g, 3.5g, 1.8g, 0.8g

Maker's Marks: none

Verification Stamps:

Adjustments: none

Form, Lid: unknown

Form, House: unknown

Decoration, Lid: unknown

Decoration, House: unknown

Dimensions: 4.2 x 3.2 x 2.8

Condition: incomplete; house and 2 cups missing

History: ECS/before 1940

Comments: Madrid is 460.0g and Castile is 460.5g. Ref. Lavagne, 1981, p. 152–53.

Catalog No.: 63

Collection No.: CIA2

Usage Type: Commercial

Century: 19th

Place of Manufacture: Spain

Place of Use: Spain; Madrid

Mass Quantity: 1 kilogramo (I KILOGRAMO)

Theoretical Mass: 1000g

Numerical System: decimal

Numerical Breakdown: 1 kilogram; 500 (500 GRAMOS), 200 (200 Gr.), 100 (100 Gr.), 100 (100 Gr.), 50 (50 Gr.), 20 (20 Gr.), 10 (10 Gr.), [10], [5], [2], [2], [1] gram

Actual Mass: 500.0g, 200.0g, 100.0g, 100.0g, 50.0g, 20.0g, 10.0g

Maker's Marks: none

Verification Stamps:

D.F. 90

Adjustments: none

Form, Lid: single-arm w/o handle

Form, House: straight; high type

Decoration, Lid: plain; the "arm" is a simple bar

Decoration, House: plain

Dimensions: 6.3 x 4.2 x 5.5

Condition: incomplete; 5 missing cups; lid does not close flush with house; hinge is bent

History: CCF/1958

Comments: Cf. Lavagne, 1981, p. 154–55.

Catalog No.: 64

Collection No.: CISw1

Usage Type: Commercial

Century: 19th

Place of Manufacture: Sweden

Place of Use: Sweden

Mass Quantity: 1 pund
(IH)

Theoretical Mass: 425.1g per Swedish pound

Numerical System: binary

Numerical Breakdown: 1 pund; (16), (8), (4), (2), (1), (1/2), [1/4], [1/8], [1/8] lod

Actual Mass: 211.4g, 106.3g, 52.8g, 26.5g, 13.2g, 6.6g

Maker's Marks: none

Verification Stamps:

Adjustments: none

Form, Lid: single-arm w/o handle

Form, House: straight; high type

Decoration, Lid: the "arm" is fluted, with cut-out designs, wider at one end and narrower at the other

Decoration, House: plain

Dimensions: 5.0 x 3.1 x 3.6

Condition: incomplete; 3 missing cups

History: ECS/1944–5

Comments: Unlike the three crown mark on Nuremberg nested weights, the Swedish three crowns are often upside down, as in this example, or in other multiples and configurations, as in this example on the inside and outside bottom of some cups. Ref. Houben, 1984, p. 36, fig. 67; cf. Numismatica Wien, 1974, no. 136.

Catalog No.: 65

Collection No.: CIRu1

Usage Type: Commercial

Century: 19th

Place of Manufacture: Russia

Place of Use: Russia

Mass Quantity: 48 solotnik/ 1/2 pound

Theoretical Mass: 409.5g per Russian pound

Numerical System: duodecimal

Numerical Breakdown: 48, (24), (12), [6], [3], [2], [1/2], [1/2] solotnik

Actual Mass: 102.4g, 51.2g

Maker's Marks: none

Verification Stamps:

18  95

пar.

Adjustments: none

Form, Lid: domed

Form, House: barrel-shaped, with convex sides; low type

Decoration, Lid: plain

Decoration, House: plain

Dimensions: 3.5 x 4.3 (diameter of base x height at greatest point)

Condition: incomplete; 5 missing cups

History: ECS/1941–42

Comments: Cf. Mönnig, 1980, 278–79 and 1983, 564; cf. Houben, 1984, p. 23, fig. 37.

Catalog No.: 66

Collection No.: CIRu2

Usage Type: Commercial

Century: 19th

Place of Manufacture: Russia

Place of Use: Russia

Mass Quantity: 48 solotnik/ 1/2 pound

Theoretical Mass: 409.5g per Russian pound

Numerical System: duodecimal

Numerical Breakdown: 48, (24), (12), {(6)}, [3], [2], [1/2], [1/2] solotnik

Actual Mass: 100.5g, 49.6g, {25.6g}

Maker's Marks: none

Verification Stamps:

Adjustments: subtractions: routing inside 12 solotnik cup

Form, Lid: domed

Form, House: barrel-shaped, with convex sides; low type

Decoration, Lid: plain

Decoration, House: plain

Dimensions: 3.5 x 5.3 (diameter of base x height at greatest point)

Condition: incomplete; 4 missing cups; 1 extrinsic cup

History: ECS/1945–46

Comments: See catalog no. 65.

Catalog No.: 67

Collection No.: CIRu3

Usage Type: Commercial

Century: early 20th

Place of Manufacture: Russia

Place of Use: Russia

Mass Quantity: 96 solotnik/1 pound

Theoretical Mass: 409.5g per Russian pound

Numerical System: duodecimal

Numerical Breakdown: 96, [48], (24), (12 **3**), (6 **3**), [3], (2 **3**), [1/2], [1/2] solotnik

Actual Mass: 102.5g, 51.3g, 25.6g, 8.1g

Maker's Marks: none

Verification Stamps:

Adjustments: subtractions: filing on bottom of 2 solotnik cup

Form, Lid: unknown

Form, House: unknown

Decoration, Lid: unknown

Decoration, House: unknown

Dimensions: 3.5 x 2.9 (diameter of base x height at greatest point)

Condition: incomplete; house and 3 cups missing

History: ECS/1943–44

Comments: See Catalog No. 65.

Catalog No.: 68

Collection No.: CIGB5

Usage Type: Bullion

Century: 19th (pre-1878)

Place of Manufacture: England

Place of Use: England

Mass Quantity: 32 oz. troy (TROY)

Theoretical Mass: 373.2g per Troy pound of 12 ounces

Numerical System: binary

Numerical Breakdown: 32, 16 (16 OZ TROY), 8 (8 OZ TROY), 4 (4 OZ TROY), 2 (2 OZ TROY), 1, 1/2, 1/4, 1/4 ounce

Actual Mass: 497.7g, 249.0g, 124.4g, 62.0g, 31.0g, 15.5g, 7.7g, 7.7g

Maker's Marks: none

Verification Stamps: none

Adjustments: subtractions: routing on bottom and sides of cups

Form, Lid: lidless

Form, House: open set

Decoration, Lid: none

Decoration, House: outer cup has incised lines; inner cups do not

Dimensions: 6.5 x 4.5 x 5.1

Condition: complete

History: ECS/before 1940

Comments: According to Crawforth, 4–21–85, the 16,8,4 binary progression of troy weights, when the troy pound is 12 ounces, was based on practicality—twelve does not divide conveniently below three. The troy pound was abolished in 1878 and thereafter the largest troy unit was the ounce. The total mass of the entire set is 996.3g.

Catalog No.: 69

Collection No.: CIGB2

Usage Type: Bullion

Century: 19th (pre-1878)

Place of Manufacture: England

Place of Use: England

Mass Quantity: 32 oz. troy

Theoretical Mass: 373.2g per troy pound of 12 ounces (see comments Cat. No. 68)

Numerical System: binary

Numerical Breakdown: 32, 16 (XVI R). 8 (VIII R), 4 (IIIIR),2 (II R), 1, 1/2, 1/4, 1/4 ounce

Actual Mass: 497.8g, 249.4g, 124.5g, 62.0g, 31.0g, 15.5g, 7.7g, 7.7g

Maker's Marks: none

Verification Stamps: (see comments Cat. No. 72)

Adjustments: subtractions: routing on bottom of cups

Form, Lid: lidless

Form, House: open set

Decoration, Lid: none

Decoration, House: none

Dimensions: 6.6 x 4.7 x 4.6

Condition: complete

History: ECS/before 1940

Comments: According to Crawforth, 8–29–86, the great majority of complete sets have incised lines around the outside of the outer cup, but the inner cups were usually plain. Few were done in the reverse, as is the case here. There seems to be little way of telling whether sets are complete or not based on the presence or absence of these lines. The total mass of the entire set is 995.5g.

Catalog No.: 70

Collection No.: CIGB7

Usage Type: Bullion

Century: 18th

Place of Manufacture: England

Place of Use: England

Mass Quantity: 16 oz. troy (TROY)

Theoretical Mass: 373.2g per troy pound of 12 ounces (see comments Cat. No. 68)

Numerical System: binary

Numerical Breakdown: 16, {8 (8 OZ TROY)}, {4 (4 OZ TROY)}, {2 (2 OZ TROY)}, {1 (1 OZ TR)}, [1/2], [1/4], [1/4] ounces

Actual Mass: {248.8g}, {124.8g}, {62.2g}, {30.5g}

Maker's Marks: none

Verification Stamps: none

Adjustments: additions: a piece of sheet copper (?) has been cut out and fitted into the bottom of the 4 oz. cup.

Form, Lid: lidless

Form, House: open set

Decoration, Lid: none

Decoration, House: incised lines on outer cup

Dimensions: 5.6 x 4.4 x 3.7

Condition: incomplete; 3 missing cups; 4 extrinsic cups

History: ECS/1943

Comments: The entire set is an "artificial" one. There are incised lines on the 8 oz. and 2 oz. cups.

Catalog No.: 71

Collection No.: CIGB6

Usage Type: Bullion

Century: 19th

Place of Manufacture: England

Place of Use: England

Mass Quantity: 16 oz. troy

Theoretical Mass: 373.2g per troy pound of 12 ounces (see comments Cat. No. 68)

Numerical System: binary

Numerical Breakdown: 16, 8 (VIII ℞), 4 (IIII ℞), 2 (II ℞), 1 (I ℞), 1/2, 1/4, 1/8, 1/16, 1/32, [1/32] ounce

Actual Mass: 249.1g, 124.7g, 62.1g, 30.9g, 15.5g, 7.7g, 3.7g, 1.8g, 0.9g

Maker's Marks: none

Verification Stamps:

Adjustments: subtractions: routing, filing and partial refilling with lead on bottom of cups

Form, Lid: flat w/o handle

Form, House: straight; high type

Decoration, Lid: plain

Decoration, House: plain

Dimensions: 5.0 x 3.4 x 4.0

Condition: incomplete; 1 missing cup

History: ECS/1947

Comments: According to Crawforth, 8–29–86, the Ewer was used by Founders' Company, i.e., The Worshipful Company (or Guild) of Founders. See also comments on Cat. No. 72. Cf. Brown, no. 115; cf. Schenk-Behrens, 1987, no. 597.

Catalog No.: 72

Collection No.: CIGB1

Usage Type: Bullion

Century: 19th

Place of Manufacture: England

Place of Use: England

Mass Quantity: 16 oz. troy

Theoretical Mass: 373.2g per troy pound of 12 ounces (see comments Cat. No. 68)

Numerical System: binary

Numerical Breakdown: 16, 8 (VIII R), 4 (IIII R), 2 (II R), 1 (I R), 1/2, 1/4, 1/4 ounce

Actual Mass: 495.6g, 246.8g, 124.8g, 62.2g, 31.0g, 15.5g, 7.7g, 7.7g

Maker's Marks: none

Verification Stamps:

Adjustments: subtractions: routing on bottom of house and cups

Form, Lid: flat w/o handle

Form, House: straight; high type

Decoration, Lid: plain

Decoration, House: plain

Dimensions: 5.1 x 3.8 x 4.0

Condition: complete

History: ECS/1947

Comments: The verification mark is from Goldsmith's Hall, London. According to Crawforth, 4–21–85, one can not deduce that these weights were used in London from the stamps because weights verified by Founders' Company or Goldsmith's Hall were used all over the country. It was a source of pride for provincial scalemakers to be able to state that they used weights verified in London. Cf. Schenk-Behrens, 1987, no. 597.

Catalog No.: 73

Collection No.: CIGB10

Usage Type: Bullion

Century: 19th

Place of Manufacture: England

Place of Use: England

Mass Quantity: 16 oz. troy

Theoretical Mass: 373.2g per troy pound of 12 ounces (see comments Cat. No. 68)

Numerical System: binary

Numerical Breakdown: 16, 8 (VIII R), 4 (IIII R), 2 (II R), 1 (I R), 1/2, [1/4], [1/4] ounce

Actual Mass: 248.9g, 124.3g, 62.2g, 31.0g, 15.6g

Maker's Marks: none

Verification Stamps:

Adjustments: subtractions: routing on bottom of house and cups

Form, Lid: flat w/o handle

Form, House: straight; high type

Decoration, Lid: plain

Decoration, House: plain

Dimensions: 4.8 x 3.1 x 4.3

Condition: incomplete; 2 missing cups

History: ECS/1947

Comments: See comments on Cat. No. 72. Cf. Schenk-Behrens, 1987, no. 597.

Catalog No.: 74

Collection No.: CIGB3

Usage Type: Bullion

Century: 19th

Place of Manufacture: England

Place of Use: England

Mass Quantity: 16 oz. troy

Theoretical Mass: 373.2g per troy pound of 12 ounces (see comments Cat. No. 68)

Numerical System: binary

Numerical Breakdown: 16, 8 (8 OZ TROY), 4 (4 OZ TROY), 2 (2 OZ), 1 (1 OZ), 1/2, 1/4, 1/4 ounce

Actual Mass: 497.8g, 248.9g, 124.4g, 62.0g, 30.9g, 15.4g, 7.6g, 7.6g

Maker's Marks: none

Verification Stamps: none

Adjustments: none

Form, Lid: lidless

Form, House: open set

Decoration, Lid: none

Decoration, House: no incised lines on outer cup, but inner 4 oz. cup has incised lines (See comments Cat. No. 69)

Dimensions: 5.4 x 3.8 x 3.7

Condition: complete

History: ECS/1941–43

Comments: The inner cups are not flush with the rim of the outer cup; they are very slightly above it. It is possible that the 8 oz. cup is extrinsic to the set, making it an 8 oz. set instead of a 16 oz. set. The mass of the entire set is 497.8g.

Catalog No.: 75

Collection No.: CIGB14

Usage Type: Bullion/Commercial

Century: 19th

Place of Manufacture: England

Place of Use: England

Mass Quantity: 16 oz. troy/ 16 oz. avoirdupois

Theoretical Mass: binary

Numerical System: 373.2g per troy pound of 12 ounces/ 453.6g per avoirdupois pound of 16 ounces

Numerical Breakdown: 16, [8], 4 (IIII R), 2 (II R), 1 (I R), 1/2, 1/4, 1/4 ounce

Actual Mass: 112.8g, 56.1g, 28.2g, 14.1g, 7.1g, 7.1g

Maker's Marks: none

Verification Stamps: none

Adjustments: subtractions: deep routing in shape of large '+' on bottom of cup

Form, Lid: lidless

Form, House: open set

Decoration, Lid: none

Decoration, House: none

Dimensions: 4.2 x 3.4 x 2.3

Condition: incomplete; 2 outer cups missing

History: unknown

Comments: This set was adjusted from troy to avoirdupois. Cf. Brown, no. 114.

Catalog No.: 76

Collection No.: CIGB9

Usage Type: Bullion

Century: 19th

Place of Manufacture: England

Place of Use: England

Mass Quantity: 8 oz. troy

Theoretical Mass: 373.2g per troy pound of 12 ounces (see comments Cat. No. 68)

Numerical System: binary

Numerical Breakdown: 8, 4 (4 OZ TROY), 2, 1, 1/2, 1/4, 1/8, [1/8] ounce

Actual Mass: 124.4g, 62.0g, 31.0g, 15.5g, 7.7g, 3.7g

Maker's Marks: none

Verification Stamps: none

Adjustments: additions: lead soldered onto bottom of cups

Form, Lid: lidless

Form, House: open set

Decoration, Lid: none

Decoration, House: incised lines on outer cups

Dimensions: 4.3 x 3.3 x 2.7

Condition: incomplete; 1 missing cup

History: ECS/1947

Comments: See comments Cat. No. 69. Cf. Uit den Boogaard, 1981, p. 292, fig. 3.

Catalog No.: 77

Collection No.: CIGB8

Usage Type: Bullion

Century: 19th

Place of Manufacture: England

Place of Use: England

Mass Quantity: 8 oz. troy

Theoretical Mass: 373.2g per troy pound of 12 ounces (see comments Cat. No. 68)

Numerical System: binary

Numerical Breakdown: 8, [4], 2 (2 TROY), (1), (1/2), (1/4), (1/4) ounce

Actual Mass: 62.3g, 31.1g, 15.5g, 7.7g, 7.7g

Maker's Marks: none

Verification Stamps: none

Adjustments: none

Form, Lid: lidless

Form, House: open set with very prominent inner weight knob (see comments Cat. No. 78)

Decoration, Lid: none

Decoration, House: none

Dimensions: 3.3 x 2.6 x 2.3

Condition: incomplete; 1 missing cup

History: unknown

Comments: The outer cup is missing. There are no incised lines on these cups.

Catalog No.: 78

Collection No.: CIGB4

Usage Type: Bullion

Century: 19th

Place of Manufacture: England

Place of Use: England

Mass Quantity: 8 oz. troy

Theoretical Mass: 373.2g per Troy pound of 12 ounces (See comments Cat. No. 68)

Numerical System: binary

Numerical Breakdown: 8, 4 (4 OZ TROY), 2 (2 OZ TROY), 1 (1 OZ), 1/2, {1/2}/ [1/4], [1/4] ounce

Actual Mass: 248.0g, 124.1g, 61.9g, 30.7g, 15.3g, {15.4g}

Maker's Marks: none

Verification Stamps: none

Adjustments: none

Form, Lid: lidless

Form, House: open set

Decoration, Lid: none

Decoration, House: incised lines on outer cup only

Dimensions: 4.3 x 3.2 x 2.6

Condition: incomplete; 2 missing cups; 1 extrinsic weight (see comments)

History: ECS/1941–43

Comments: According to Crawforth, 8–29–86, the innermost 1/2 oz. weight, which looks like two 1/4 oz. weights stuck together, is actually a form of flat-knobbed weight (for picking up with the fingernail) and is apparently a British idea. Cf. Uit den Boogaard, p. 292, fig. 3.

Catalog No.: 79

Collection No.: CIGB13

Usage Type: Commercial

Century: 19th

Place of Manufacture: England

Place of Use: England

Mass Quantity: 16 oz. avoirdupois

Theoretical Mass: 453.6g per 1 avoirdupois pound of 16 ounces

Numerical System: binary

Numerical Breakdown: 16, 8 (8 oz), 4 (4 oz), {2 (II)}, {1 (I)}, [1/2], [1/4], [1/4] ounce

Actual Mass: 226.8g, 113.4g, {60.5g}, {29.7g}

Maker's Marks: none

Verification Stamps: (see comments Cat. No. 72)

Adjustments: subtractions: routing

Form, Lid: lidless

Form, House: open set

Decoration, Lid: none

Decoration, House: incised lines on outer cup

Dimensions: 5.4 x 4.0 x 3.5

Condition: incomplete; 3 missing cups; 2 extrinsic cups

History: ECS/1943

Comments: There are incised lines on all but the 4 oz. cup. Cf. Houben, 1984, p. 20, fig. 29.

Catalog No.: 80

Collection No.: CIUS1

Usage Type: Inspector's set

Century: early 20th

Place of Manufacture: U.S.A.

Place of Use: United States

Mass Quantity: 8 pounds avoirdupois

Theoretical Mass: 453.6g per avoirdupois pound

Numerical System: binary

Numerical Breakdown: 8, 4 (4 LB. AVD.), (2), (1) pound; 8 (8 OZ.), (4), (2), (1), (1/2), (1/4), (1/4) ounce

Actual Mass: 1813.5g, 907.2g, 453.6g, 226.4g, 113.5g, 56.6g, 28.3g, 14.1g, 7.1g, 7.1g

Maker's Marks: W. & L. E. Gurley (this is stamped into the rim of the largest cup)

Verification Stamps:

N.Y. 08
N.Y. 10

FC 31 C1 13 23

Adjustments: additions: lead soldered into the bottoms of the smallest cups; screws in the bottoms of the largest cups

Form, Lid: lidless

Form, House: open set; stitched leather case (see comments)

Decoration, Lid: none

Decoration, House: none

Dimensions: 10.3 x 8.1 x 6.9 (without case); 11.7 x 8.7 x 8.0 (case)

Condition: complete; leather case is beginning to react with and corrode the outer cup

History: ECS/before 1940

Comments: This set is contained in a tightly fitted, stitched leather case with a strap and buckle closure. Cf. Gurley, 1912, p. 86, nos. 9400 and 9402.

BIBLIOGRAPHY

Batz, G. Personal communication (visit to collection), 20–21 October 1986.

______. Letter to author, 11 November 1986.

Brown, O. *Catalogue 2: Balances & Weights.* Cambridge: Whipple Museum of the History of Science, 1982.

Conservatoire National des Arts et Métiers. "Poids." In *Cataloge du Musée: Poids et mesures métrologie,* 93–134. Paris: Conservatoire national des arts et Métiers, 1941.

Crawforth, M. *Handbook of Old Weighing Instruments.* Chicago: International Society of Antique Scale Collectors, 1984.

______. Letter to author, 21 April 1985.

______. Letter to author, 29 August 1986.

Danforth, E. Z. "The Streeter Collection of Weights and Measures: A New Resource for Metrologists." Paper presented at the Historical Metrology Symposium, XVIIth International Congress of History of Science, University of California, Berkeley, August 1985.

Doursther, H. *Dictionnaire universel des poids et mesures anciens et modernes.* 1840. Reprint. Amsterdam: Meriden Publishing Co., 1965.

Forien de Rochesnard, J. *Dictionnaire ponderal.* Vols. 1 and 2. Paris: Private printing, 1967.

Forien de Rochesnard, J., and F. Lavagne, "Les poids d'Alsace et de Lorraine." *Congrés national des sociétés savantes,* Strasbourg, 1967, archéologie, 67–121. Paris: Bibliothèque National, 1970.

Gurley, W. & L. E. *A Handbook for the Use of Sealers of Weights and Measures.* 4th ed. Troy, New York: W. & L. E. Gurley, 1912.

Houben, G. M. M. *The Weighing of Money.* Zwolle: G. M. M. Houben, 1982.

______. Letter to the Historical Library, 24 June 1983.

______. *2000 Years of Nested Cup-Weights.* Zwolle: G. M. M. Houben, 1984.

______. Letter to author, 28 May 1985.

______. Letter to author, 15 July 1986.

Kisch, B. Z. *Gewichte- und Waagemacher im alten Köln*. Cologne: Verlag der Lówe, Dr. Hans Reykers, 1960.

______. *Scales and Weights: An Historical Outline*. New Haven: Yale University Press, 1965.

Lavagne, F. "Poids a godets pour pesage monétaire." *Gazette Numismatique Suisse* 18(1968): 39–47.

______. Collection Notes. Unpublished material, Streeter Collection of Weights & Measures, Yale University Medical Historical Library, New Haven.

______. *Balanciers etalonneurs: Leurs marques—leurs poinçons*. Montpelliers: 1981.

Lockner, H. P. *Die Merkzeichen der Nürnberger Rotschmiede*. Munich: Deutscher Kunstverlag, 1981.

Mönnig, C. "Russian Nesting Weights." *Equilibrium* 3(1980):278-79.

______. "Cyrillic Letter." *Equilibrium* 6(1983):564.

"Notes & Queries." *Equilibrium* 1(1978): 16–17.

Numismatica Wein. *Auktion VI: Munz-waagen, Munz-gewichte*. Wein: 1974.

Schenk-Behrens, K. W. *Essener Waagen-Auktion I*. Essen: Karla W. Schenk-Behrens, 1981.

______. *Essener Waagen-Auktion II*. Essen: Karla W. Schenk-Behrens, 1982.

______. *Essener Waagen-Auktion III*. Essen: Karla W. Schenk-Behrens, 1984.

______. *Essener Waagen-Auktion IV*. Essen: Karla W. Schenk-Behrens, 1985.

______. *Essener Waagen-Auktion V*. Essen: Karla W. Schenk-Behrens, 1987.

Skinner, F. G. *Weights and Measures: Their Origins and their Development in Great Britian up to A. D. 1855*. London: HMSO, 1967.

Stengel, W. "Die Merkzeichen der Nürnberger Rotschmiede." In *Mitteilungen aus dem germanischen Nationalmuseum*, 107–55. Nuremberg: Nürnberg Germanisches National Museum, 1918.

Streeter, E. C. "Comment on Collecting." Talk presented before the Travel Club, New Haven, Connecticut, 26 November 1940. Unpublished manuscript, E. C. Streeter Collection of Weights & Measures Yale University Medical Historical Library, New Haven.

Streeter, J. W. "The Edward Clark Streeter Collection of Weights and Measures at the Yale Medical Library—One of Yale's Least Known Bests." *Yale University Staff News* 6(1974):4-5.

Uit den Boogaard, L. A. "1001 Variations." *Equilibrium* 4(1981):291–98.

Yale University. *The Making of a Library: Extracts from Letters 1934–1941 of Harvey Cushing, Arnold C. Klebs, John F. Fulton.* New Haven: The Carl Purington Rollings Printing Office of the Yale University Press, 1959.

Zupko, R. E. *A Dictionary of English Weights and Measures From Anglo-Saxon Times to the Nineteenth Century.* Madison: The University of Wisconsin Press, 1968.

———. *French Weights and Measures Before the Revolution: A dictionary of Provincial and Local Units.* Bloomington: Indiana University Press, 1978.

———. *Italian Weights and Measures From the Middle Ages to the Nineteenth Century.* Philadelphia: American Philosophical Society, 1985.

———. *A Dictionary of Weights and Measures for the British Isles: The Middle Ages to the Twentieth Century.* Philadelphia: American Philosophical Society, 1985.

INDEX TO MASS QUANTITY AND UNIT NAMES BY LANGUAGE

Note: The English list includes all the catalog entries. Catalog numbers, not page numbers are used.

English

French

INDEX TO THE ACTUAL MASS OF SINGLE UNITS

Note: If the single unit weight is missing from the set, the actual mass of the next largest weight (+) or the next smallest weight (−) is given instead. This index does not include the metric units.

*denotes a 12-ounce pound

This book was composed in Baskerville, an English
Transitional typeface designed by John Baskerville
during the mid-1700s.

Composed by Delmas Typesetting, Inc.
in Ann Arbor, Michigan.

Printed and bound by Thomson-Shore, Inc.
in Dexter, Michigan.

Designed by Nora Jacobson.